METAL WORKING TOOLS

and their uses

IAN BRADLEY

Model and Allied Publications

Argus Books Ltd
14 St James Road, Watford, Herts.

Model and Allied Publications,
Argus Books Ltd
14 St James Road, Watford
Herts, England

ISBN 0 85242 599 6

First published 1978

Printed by Blackfriars Press Ltd., Leicester LE5 4BS

CONTENTS

PREFACE

THE TITLE OF this book was coined many years ago by the late Percival Marshall who had perceived, as the result of enquiries appearing in his Journal *The Model Engineer,* that there was need for some simple advice aimed at those who, as he himself stated: 'were about to make their acquaintance with metal working tools for the first time'.

The original booklet, No. 7 of *The Model Engineer* series, was written nearly eighty years ago. Since then, of course, many changes have taken place and some of the many tools described and illustrated have vanished from the market.

It was clear, therefore, that to have been of immediate use, generally, the subject of the book needed to be completely rewritten using illustrations of modern equipment where needed.

In the present book the emphasis is largely on hand tools, since space would clearly inhibit adequate coverage of even the simplest of machine tools; in any case, they are well catered for elsewhere.

My grateful thanks must be extended to those firms who have provided illustrative and other material for the present book, in particular: Messrs. Black & Decker, Calor Gas and James Neill & Co who have given of their knowledge and time to this end.

HUNGERFORD: 1978   I.B.

CHAPTER I

EQUIPMENT FOR MEASURING AND MARKING OFF

IN ALL ENGINEERING production, whether they be parts or complete assemblies, it is essential that a basic source of reference should be available in order that workers may have a means of gauging the size of any parts they may be fabricating.

Rules

The carpenter has his boxwood rule, generally 2 feet in length and capable of folding for convenience sake. However well-made such a rule may be and however sharp its engraving, its use amongst the oil and grease of the metal-working shop is not a practical proposition. So instead, the machine-engraved steel rule depicted in *Fig. 1* has become the standard of reference for all metal workers.

Rules of this type are available in a number of sizes and in lengths from 6 inches to 2 feet subdivided into fractions of ½, ¼, ⅛, 1/16 and 1/32 of an inch. In addition most rules are further divided into measurements of 1/64 of an inch, which is probably the finest measurement that can be detected by the naked eye with any degree of exactness.

Rules are also produced with divisions based on decimals of an inch as well as centimetres and millimetres. In the decimal notation the divisions are 1/10, 1/20, 1/50 and 1/100 of an inch. In practice measurements smaller than 1/50 or 1/64 of an inch need to be taken either with a vernier slide gauge or a micrometer which will be described later.

As to what divisions he will need on his rule in the first instance the worker must be guided by the work he is asked to carry out. Later, perhaps, he will add rules giving him a greater range of activity.

A good quality steel rule can often be used as a straight-edge to check the flatness of any surfaces that may have been produced by filing or machining. The normal method is to apply the edge of the rule to the surface to be tested, the rule itself being tilted so that only the corner of the edge is in contact; then, by holding the work up to the light, any discrepancy will be revealed. See *Fig. 2*.

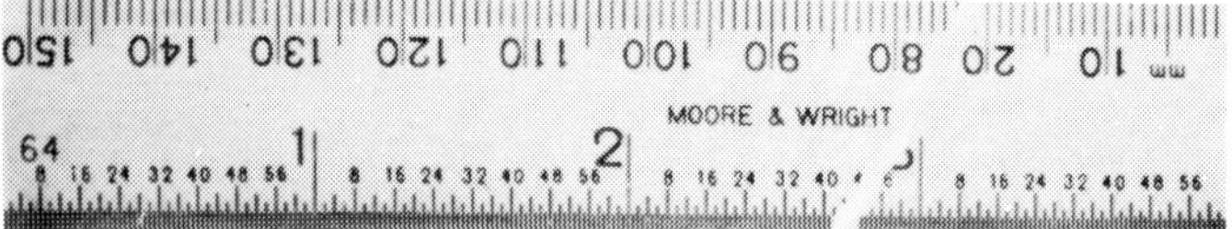

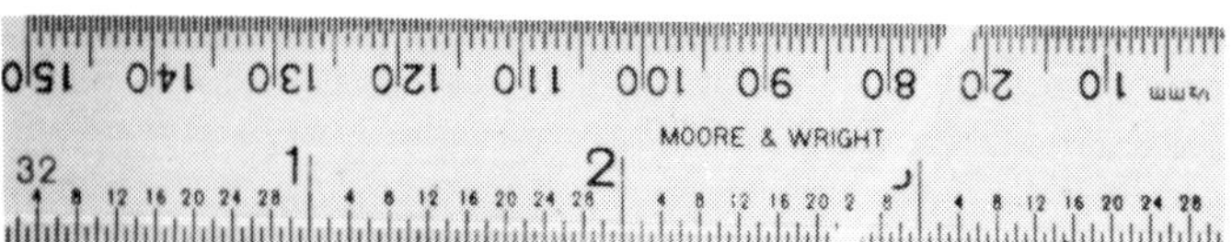

Fig. 1a

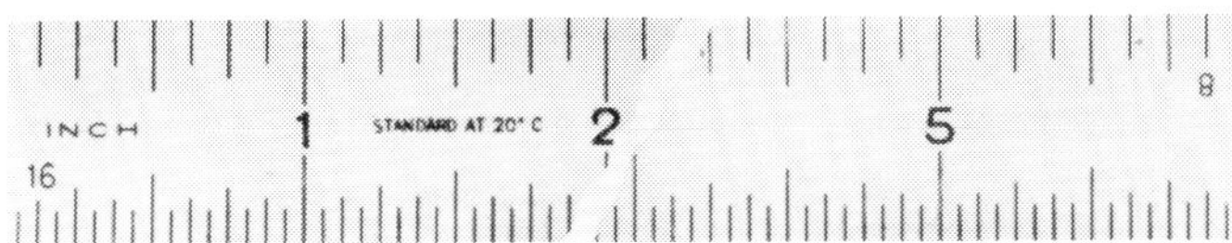

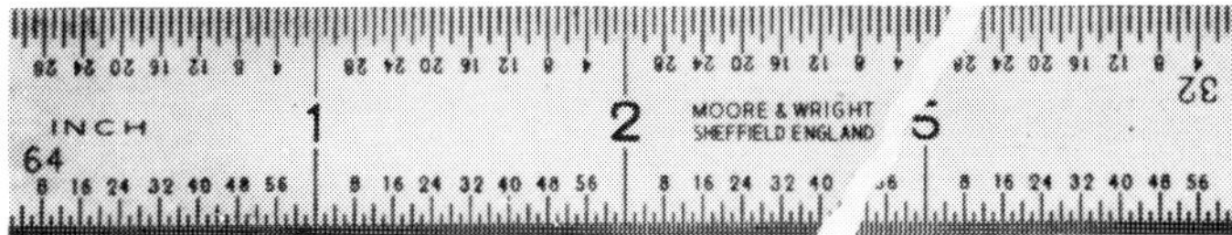

Fig. 1b

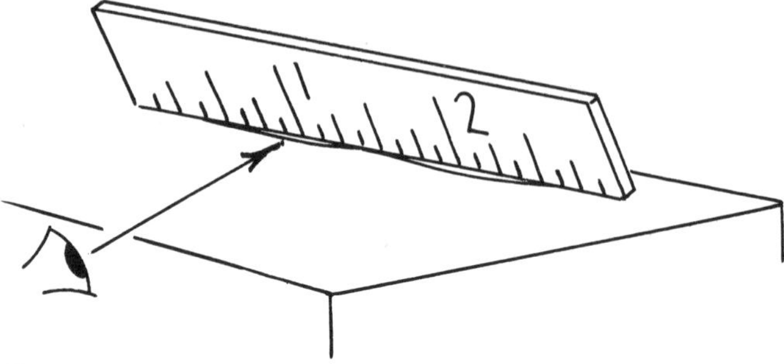

Fig. 2

The Slide Gauge

The measuring device illustrated in *Fig. 3* is known as a slide gauge, sometimes called a calliper gauge. It consists of a steel rule having at one end a short fixed jaw A protecting at right angles to the rule. The sliding jaw B has its inner surface parallel to the inner surface of the jaw A. A sleeve C is attached to the sliding jaw and is provided with a fixing screw D enabling the sleeve to be locked to the rule when needed.

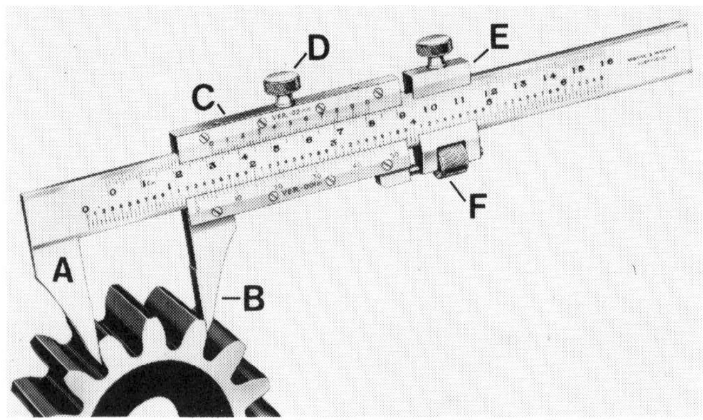

Fig. 3

A second sleeve E is also attached to the rule and is again provided with a fixing screw so that it, too, can be secured when required. This sleeve has also a fine adjustment screw F enabling a critical setting of the sliding jaw B to be made.

The sliding jaw also carries a vernier scale, enabling the user to make measurement readings to 1/1000 part of an inch or to 1/1000 of a millimetre since, for the most part, slide gauges are engraved in both inch-fractional and metric divisions.

In the example illustrated in *Fig. 4* no fine adjustment device is fitted; in many cases, however, means of fine adjustment form part of the standard slide gauge as supplied by the manufacturer.

In use the jaws of the gauge are set to required distance apart by sliding the jaw B along the scale until the edge of its sleeve is opposite the corresponding division on the rule. The sleeve C is then secured by its fixing screw D, leaving the gauge ready for use.

Conversely, when measuring a piece of work, the jaw B is moved into contact with it so that the work is gripped lightly between the jaws of the gauge. The measurement is then read off directly on the scale.

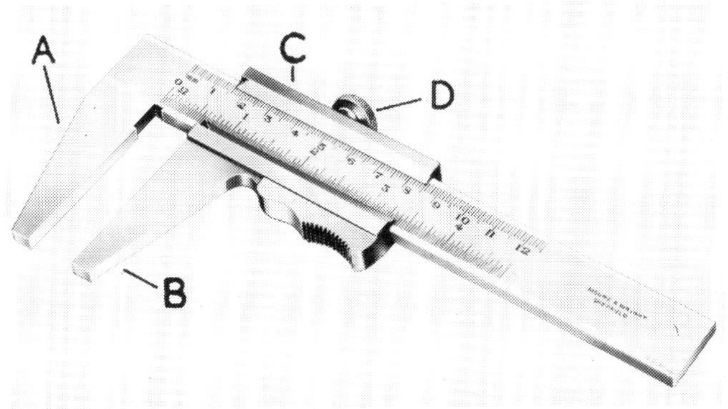

Fig. 4

The Micrometer Gauge

The micrometer screw gauge, illustrated in *Fig. 5*, was invented by Sir Joseph Whitworth in 1840. The instrument depicted is the direct descendant of Whitworth's micrometer which was in truth a very substantial machine, suitable only for mounting on a bench. Its descendant, on the other hand, is portable in every sense of the word, even in the largest sizes. As will be seen from the illustration, which shows a metric version, it comprises a U-shaped frame A in which is set an anvil B. The frame is extended at one side to form a spigot upon which are engraved divisions. On an imperial measure instrument each division is equal to 25/1000 of an inch. The spigot itself is threaded internally, 40 threads to the inch. Consequently when the spindle C, which engages this thread, is turned through one full revolution it will move axially 25/1000 of an inch, that is 1/40 inch.

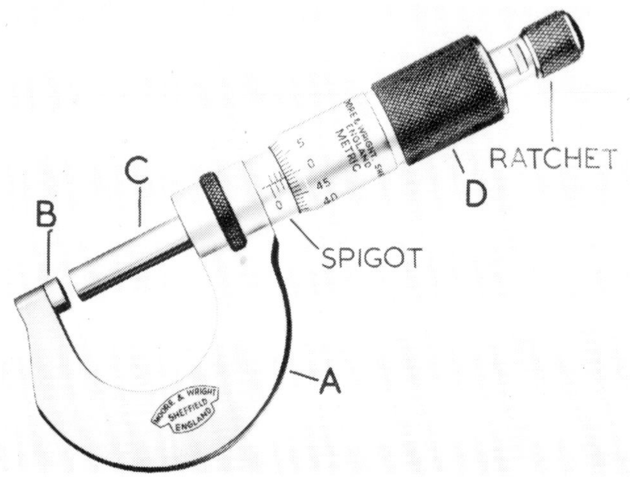

Fig. 5

At its outer end the spindle carries a sleeve or thimble D, upon which are engraved 25 equal parts. Consequently a rotation of the sleeve equal to one of these parts will result in an axial movement of the spindle amounting to 1/25 of 1/40, i.e. 1/1000 of an inch in or out according to the direction in which the sleeve is turned.

Micrometer callipers are, of course, made in many sizes. But 1-inch and 2-inch micrometers for the most part will serve well around the workshop; leaving measurements above 2 inches to be handled by vernier slide gauges well fitted for the purpose, since their range is far greater. In this connection it should perhaps be pointed out that a single 12-inch vernier slide gauge will handle a range of measurement that would otherwise need twelve individual micrometers to encompass. The range of movement of any one micrometer is 1 inch only, therefore a battery of them would be needed to cover work from 0–12 inches in diameter or length.

One should perhaps qualify these remarks a little. The slide gauge, whilst it can be used for taking measurements axially to the limit of its capacity, cannot be relied on to make such diametrical measurements unless the gauge can be supplied across the end of the work.

In the past there have been many attempts to reduce the number of micrometers needed by fitting extension devices over the anvil, indeed some combination instruments may still be marketed. But their accuracy is easily impaired.

When a number of small parts need to be measured it is very convenient to provide a stand in which the micrometer can be mounted. In this way the hands are left free to hold the work and manipulate the micrometer itself. A typical stand made by the author is illustrated in *Fig. 6*.

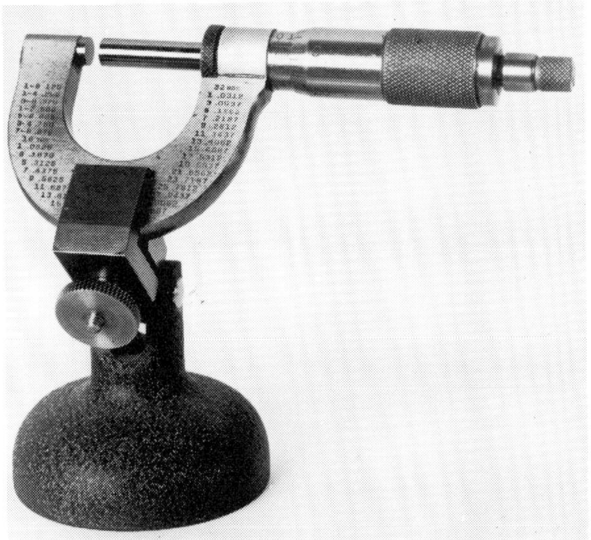

Fig. 6

Depth Gauges

It is very often necessary to measure accurately the depth of a hole or a recess in a component. The gauges used for the purpose sometimes take the form of the vernier slide gauge to which has been added a depthing pin attached to its sliding jaw. Alternatively the gauge can comprise a soleplate fitted with a vernier with a narrow 6-inch rule projecting from the soleplate. Depth gauges of this type are quick to use.

The depth gauge illustrated in *Fig. 7* is based on the micrometer already described. The particular example depicted is for metric measurement and is provided with three depth rods giving ranges: 0–24 mm,

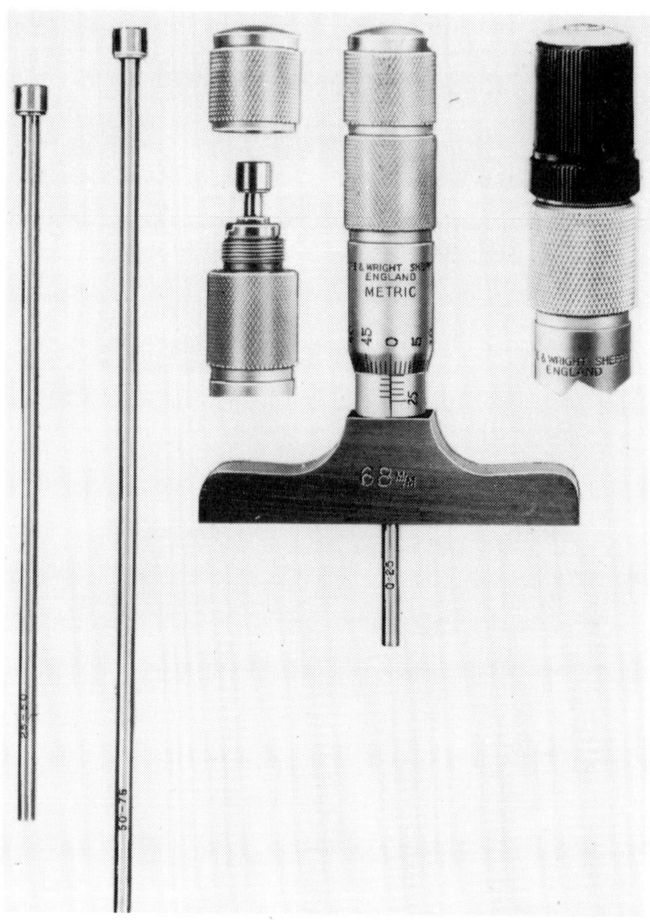

Fig. 7

25–50 mm and 50–75 mm. These pass through the spindle and are held in place by the cap seen to the left of the thimble, and for this reason, perhaps, these composite instruments have not had general acclaim.

An alternative cap is depicted to the right of the thimble. As well as securing the depth rods this cap also contains a ratchet device which some workers find convenient since it ensures that no undue tension is imparted when making a measurement. This is an important matter, for the soleplate can easily be lifted if there is lack of sensitivity when turning the thimble.

Callipers

Before the introduction of the micrometer calliper, however, the worker relied on the type of calliper illustrated in *Fig. 8* and *Fig. 9*.

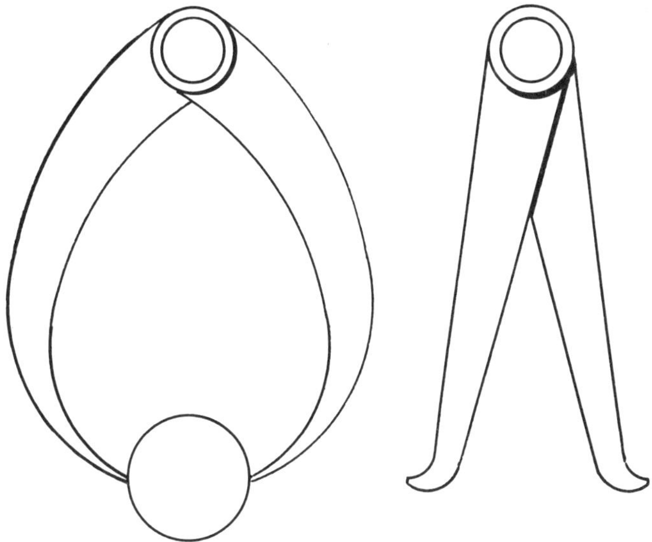

Fig. 8

Fig. 9

These are known as firm-joint callipers in that both legs form a hinge and are restrained from unwanted movement by friction only. The outside calliper seen in the first illustration is used for ascertaining the diameter of a piece of work. In use the legs are set to span the work in hand, then one leg is set against the end of a rule while the measurement is read off from the position the other leg occupies on the rule. The method is depicted in *Fig. 10*. The inside calliper depicted in *Fig. 9* is used for taking internal measurements and much the same method is employed for reading off the dimension required; but, of course, it is not possible to rest one leg of the calliper against the end of the rule.

Inside callipers are therefore set dimensionally by the method depicted in the illustration *Fig. 11*.

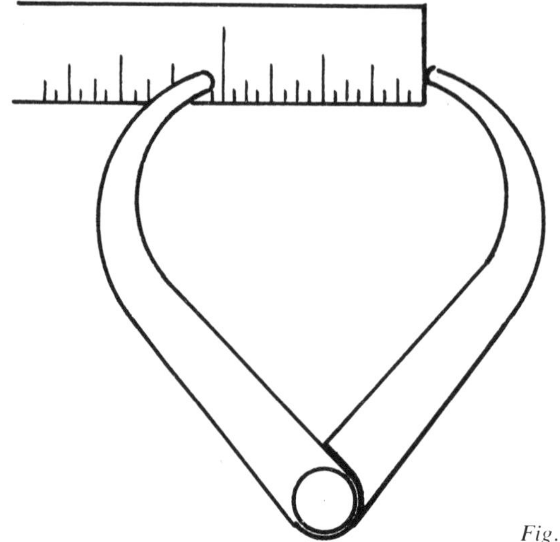

Fig. 10

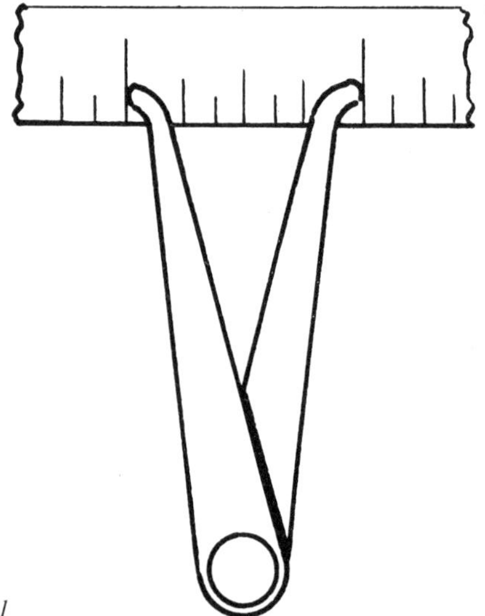

Fig. 11

Odd-leg Callipers

The callipers illustrated in *Fig. 12 and 13* are variously called 'Jennies', 'mophs', 'hermaphrodite' and 'odd-legs callipers'. They are also sometimes termed "scribing callipers" and it is for just this purpose they are intended. Their construction is similar to that of any other firm-joint calliper; but, unlike ordinary callipers where the legs are left soft, the scribing leg of the Jenny calliper is hardened to resist wear.

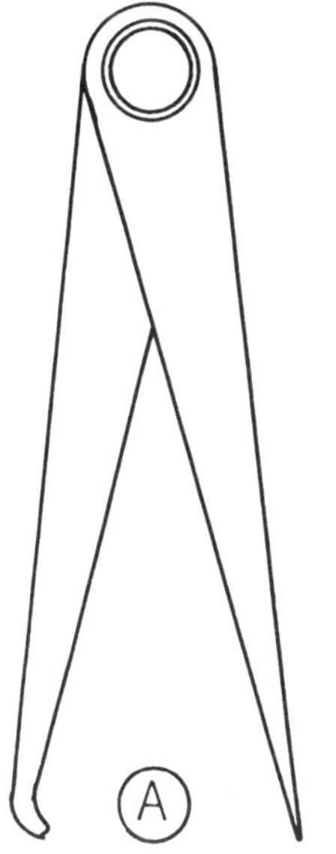

Fig. 12

Alternatively, as depicted also in the illustration, odd-legs callipers are sometimes fitted with a removable needle point; which, when hardened, is somewhat more durable, and is the more easily sharpened. A typical use of the tool is seen in the illustration *Fig. 14*.

The method of applying the tool will be apparent from the illustrations; in particular in use for finding the centre of a shaft as depicted in *Fig. 15* where, by striking a series of arcs, the central point on the end of the shaft is revealed.

Fig. 13

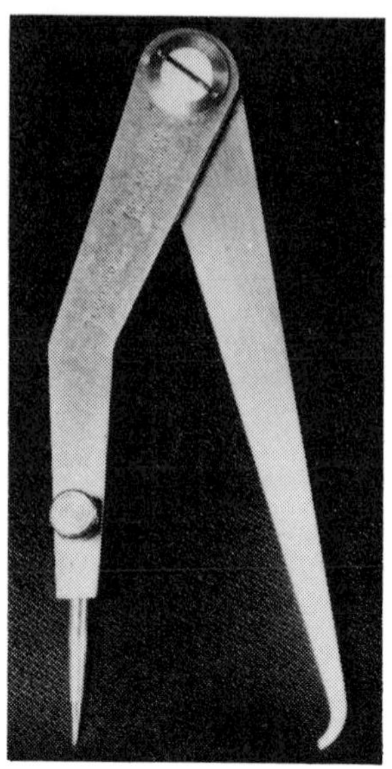

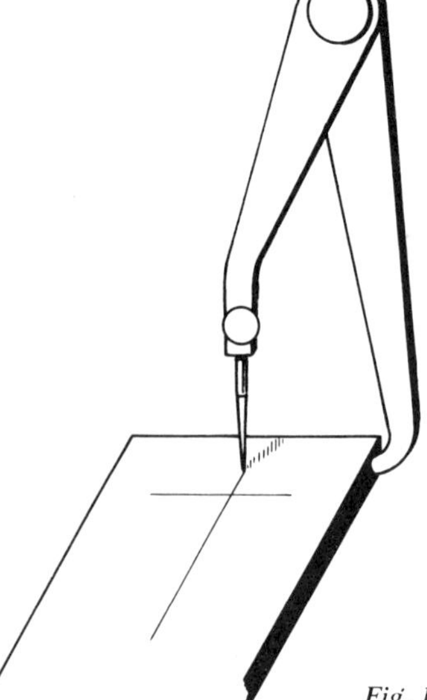

Fig. 14

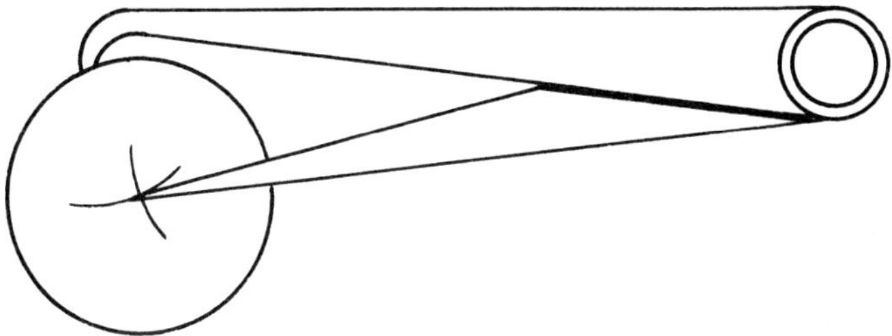

Fig. 15

Compasses or Dividers

The illustration *Fig. 16* shows a wing compass once popular among metal workers and still to be found in shops specializing in the production of large sheet metal components. In the smaller workshop, however, compasses of this type have given place to the spring dividers seen in *Fig. 17*. Indeed, callipers themselves tend to follow this form of construction which permits a far greater sensitivity than that obtainable from the firm-joint type.

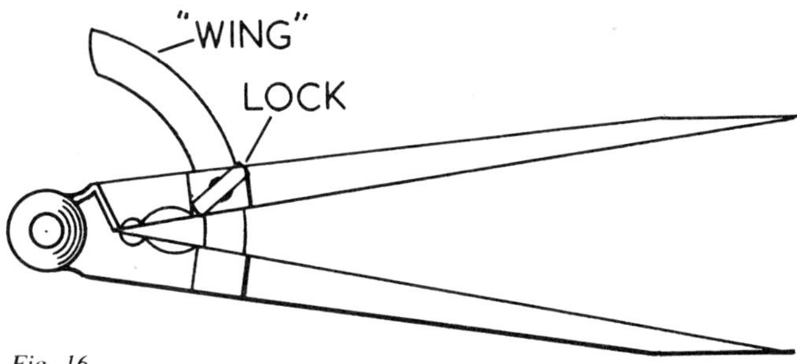

Fig. 16

Spring dividers, and the spring callipers illustrated in *Fig. 18*, have a spring joint at the top; this tends to keep the legs apart. Closure or adjustment of the legs position is effected by the milled nut and the screw seen projecting from the side of the instrument. Whilst this is naturally a slower process than that applicable to the firm-joint calliper, it is infinitely more sensitive, so spring-joint instruments are preferable for use in shops devoted to the production of components needing great accuracy.

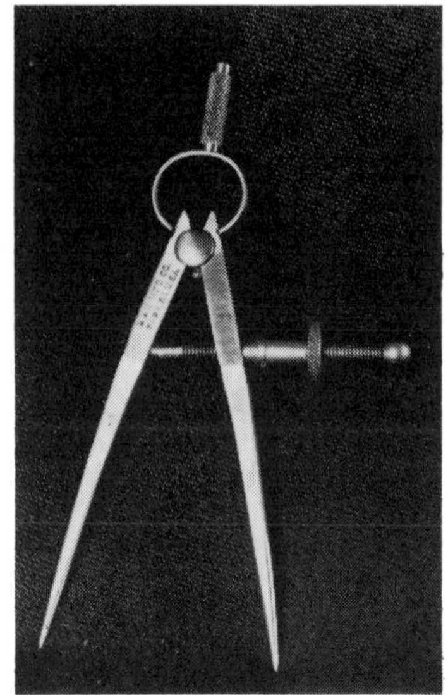

Fig. 17

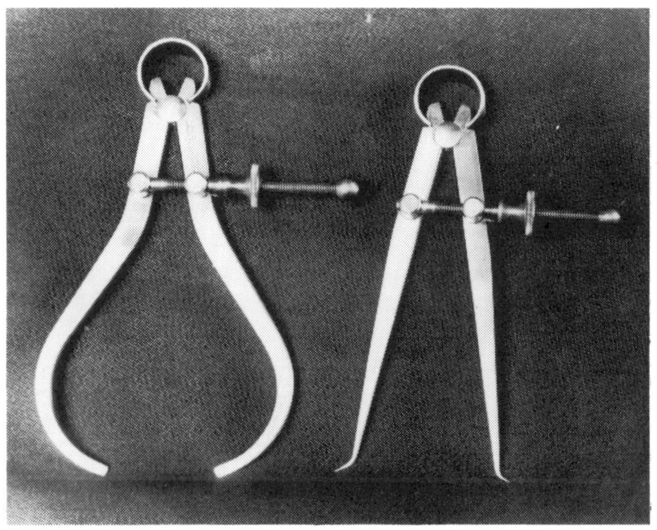

Fig. 18

The spring divider will be seen to be provided with a knurled finger grip on top of its spring. Without this the user would be forced to grip the divider by the legs, and so possibly impair the instrument's effectiveness.

The Square

The Square, depicted in *Fig. 19 and 20* is one of the most important tools in the armoury of the metal worker. By its use he can check the accuracy of any work with the file that has been performed in the bench vice; with the square he can check the uprightness of the work's setting in the drilling machine or other machine tool such as the lathe or the shaping machine.

Squares, for the most part, consist of a thin blade set truly upright in a somewhat heavier base. As may possibly be seen, the corner, where the blade joins the base, is relieved so that the square itself will fit snugly over the work. Three types of square are seen in the illustration. That on the left is the common square, that on the right is one in which the blade is also a steel rule. Such a square is useful on the marking-off table or surface plate, as may be seen later.

The third square is a somewhat specialized device. This is the toolmaker's square where the blade can be adjusted for stand-out and locked in any desired position. This is a precision tool, not for general use, and therefore expensive.

Fig. 19

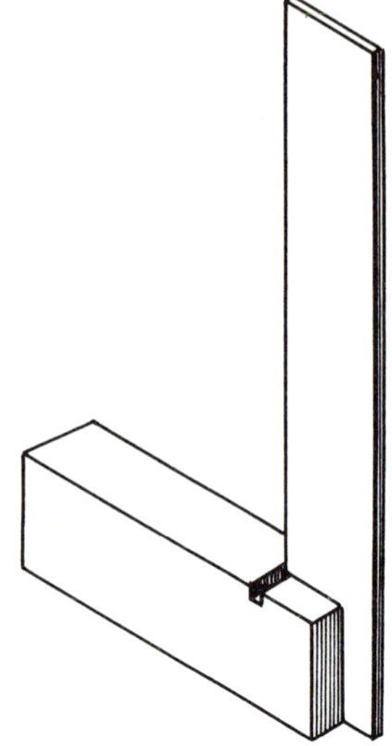

Fig. 20

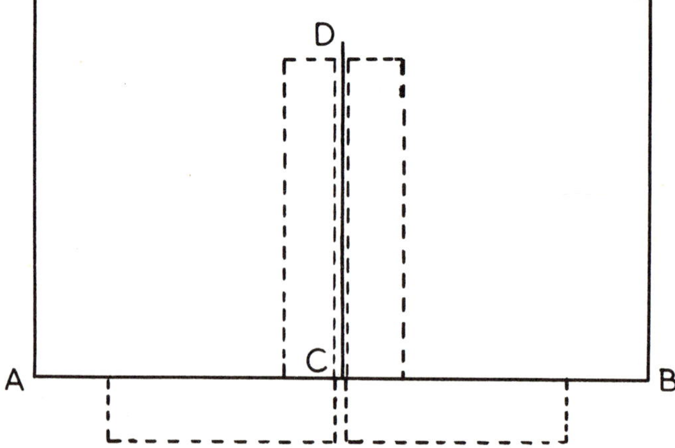

Fig. 21

Squares made commercially can usually be relied on for accuracy. However, if any doubt about a particular square exists, a simple check will reveal the fault if any. The method is depicted in the diagram, *Fig. 21*. The line AB represents an accurately planed edge on a piece of heavy gauge metal. The face of the metal plate is painted with blue marking fluid and the square to be tested is applied to the edge of the plate. A line CD is then scribed using the blade of the square, on the face of the plate. As shown by the dotted lines the square is now reversed and the blade applied to the scribed line; if the edge of the blade coincides with the scribed line exactly the square is accurate.

The inner edge of a square should be exactly parallel with the outer edge. Therefore the same test as that already described can be applied with advantage.

Combination Squares and Protractors

It is sometimes necessary to take measurements on work at angles other than 90 degrees, the only figure for which an ordinary square is

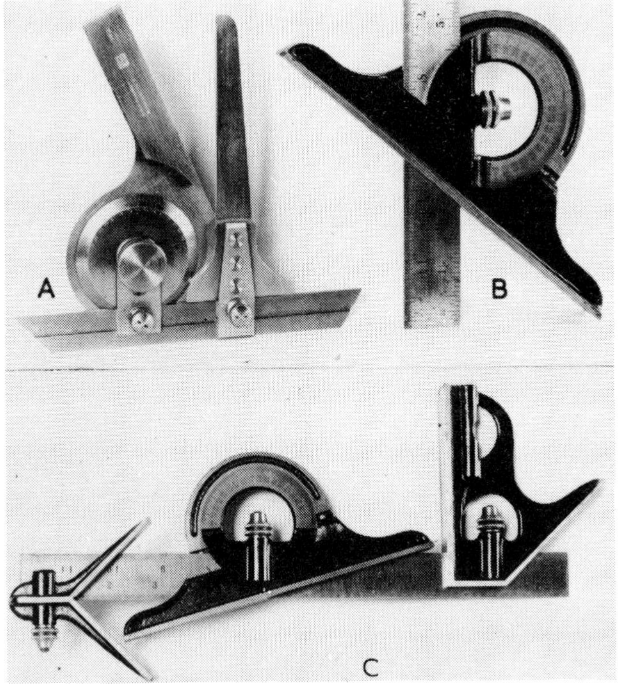

Fig. 22

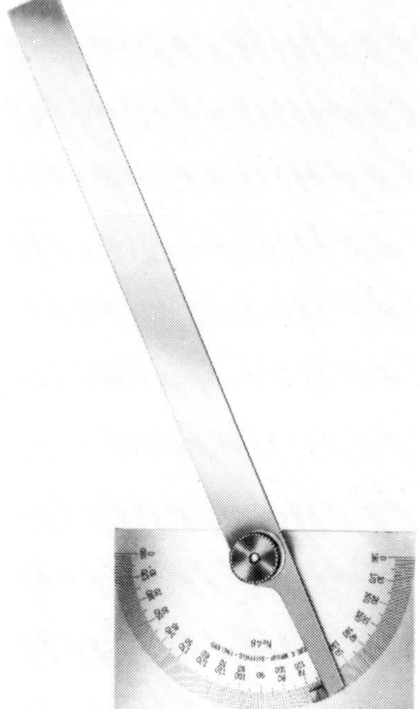

Fig. 23

suitable. In all other cases a combination square or a protractor will need to be used. The former, illustrated in *Fig. 22A, B and C,* consists in a soleplate fitted with a protractor head in which is clamped a rigid steel rule. The protractor head may be turned to bring the rule to any desired angle in relation to the soleplate. When the angle has been set, the protractor head is locked and the rule itself adjusted for position before the soleplate is applied to the work.

A simpler device is depicted in *Fig. 23.* This is an accurately made plate in which the sides are at right angles to one another. The face of the plate is engraved with a scale of degrees ranging from a central zero to 90 degrees on each side of it. An arm is pivoted at the central point of the scale and this may be locked by a knurled thumb screw at any point desired. Protractors of this type are useful when setting out work or for estimating the angularity of simple components.

The Box Square

This tool, which is sometimes called a keyseat rule, is primarily used when marking out the position of keyseats on shafting or the location of drill holes for collars or taper pins in similar components. Its use will be clear from the illustration *Fig. 24*.

The modern variant of the box square comprises a pair of accurately made clamps in which a short rule may be clamped firmly and applied to

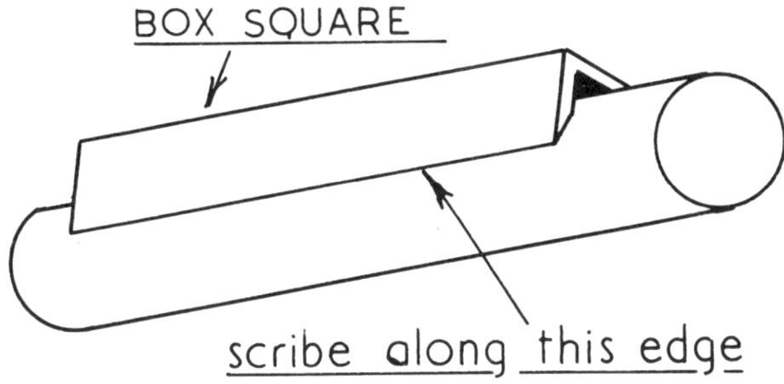

Fig. 24

the work in a manner similar to the original box square. The arrangement is depicted in the illustration *Fig. 25* from which it will no doubt be apparent that the direct use of a steel rule in this way has advantages.

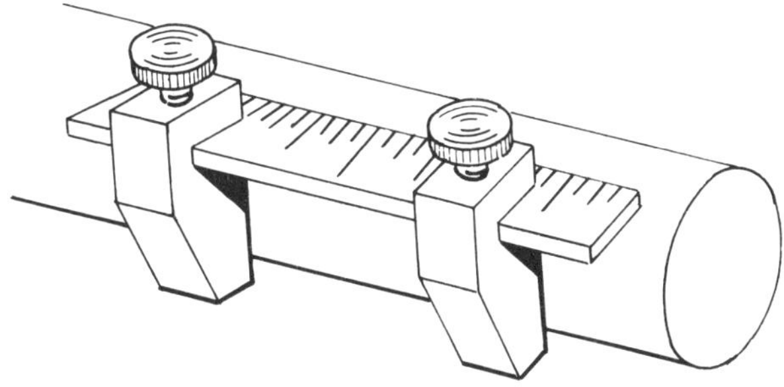

Fig. 25

The Scriber

The marking of dimension lines, or lines that trace the outline of a component, is carried out with a scriber. This may take the form of a piece of silver steel, hardened at one end to resist wear and sometimes knurled to provide an adequate finger grip. Such a scriber is illustrated at A in *Fig. 26*. An alternative form preferred by some mechanics is depicted at B. Here the scriber point is detachable so that it may be removed, reversed and placed inside the handle as a safety measure should it be necessary to put the scriber into one's pocket.

Fig. 26

The Surface Gauge

The two scribers illustrated are suitable for use on small components and general bench work; for large components, however, needing more complex treatment the scriber is usually mounted in a stand called a scribing block or surface gauge. In its simplest form this consists of a metal base A, the underside of which has been machined or filed perfectly flat, and a pillar B that carries the clamp C for holding the scriber D. The arrangement is illustrated in *Fig. 27*.

It will be appreciated that the simple surface gauge illustrated is only capable of what may loosely be termed 'rough adjustment'. For many purposes this will suffice; but many workers need a more sensitive instrument which can be used both on the bench and around machines.

On all surface gauges one end of the scriber is curved. This enables the user to employ the gauge to check the squareness of work, when set on the drilling machine or elsewhere, by applying the curved end to the work at several points on its surface.

The surface gauge depicted in *Fig. 28* is provided with a fine adjustment device consisting of a rocking lever supporting the column to which the pillar is attached and a fine adjustment screw that enables the lever to be tipped backwards and forwards against a strong spring set in

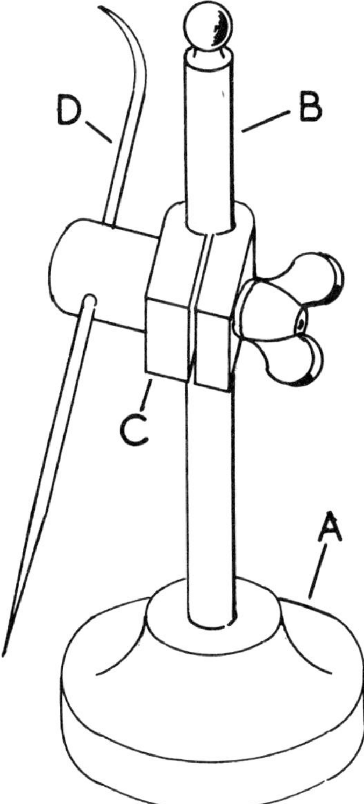

Fig. 27

the base of the gauge. The base itself is machined with a V-groove on its underside enabling it to be set on round surfaces such as shafting and the like.

The base is also provided with a pair of register pins that may be set to project from the base when required, in order that the surface gauge may be used against a reference surface such as the edge of a lathe bed for example.

A typical example of the use of the simple surface gauge is seen in *Fig. 29*. Here the casting for a pair of bearing brasses is seen being marked off before being cut in two.

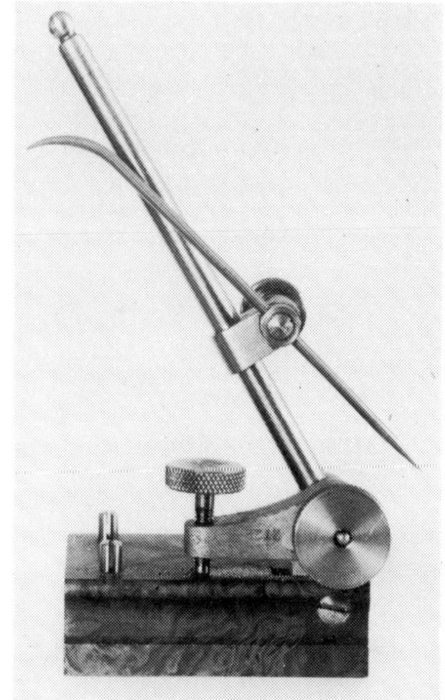

Fig. 28

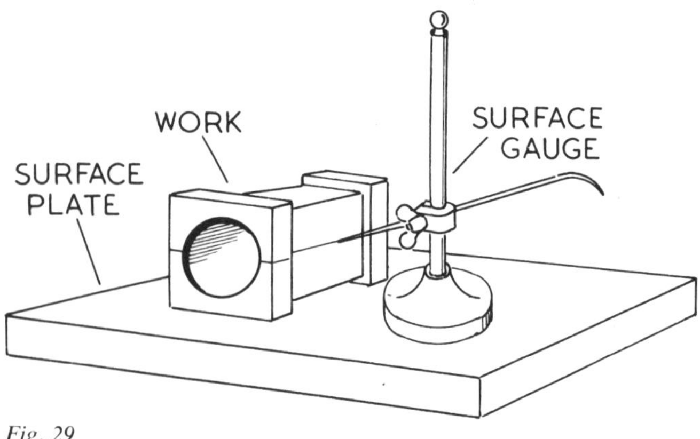

Fig. 29

The V-Block

When round material needs to be drilled or marked-off it is mounted in a
V-block or perhaps a pair of V-blocks, for the purpose. The illustration
Fig. 30 shows such an operation in progress. The V-block in use is one
of a precision matched pair whilst the operation itself is the marking-off
of a steel collar before forming an internal square.

Fig. 30

The V-block illustrated in *Fig. 31* is a cast iron block of a type
commonly used on the marking-off table. The illustration also depicts a
convenient method of holding a short rule when setting the point of the
scriber fitted to a surface gauge on the marking-off table.

Fig. 31

Centre Punches

When marking out, and after the outline and centre lines have been established, it is essential to ensure that these lines have some durability. This is achieved by making a series of punch marks on the lines which will remain in evidence even after the lines themselves may have been obliterated in working.

The sketch *Fig. 32* shows the outline and centre lines of a rough forging for a component such as a crank web. In addition the two circles indicate to the workman the size to which the seatings for the crankpin and main shaft need to be bored out.

The punches used for making the punch marks are made from cast steel and are hardened and tempered in order to reduce wear to a minimum.

For the most part, the shanks of these punches are knurled in order to provide a grip, and the points themselves are ground to an included angle of either 60° or 90° according to requirements. A typical example of a centre punch is illustrated in *Fig. 33.*

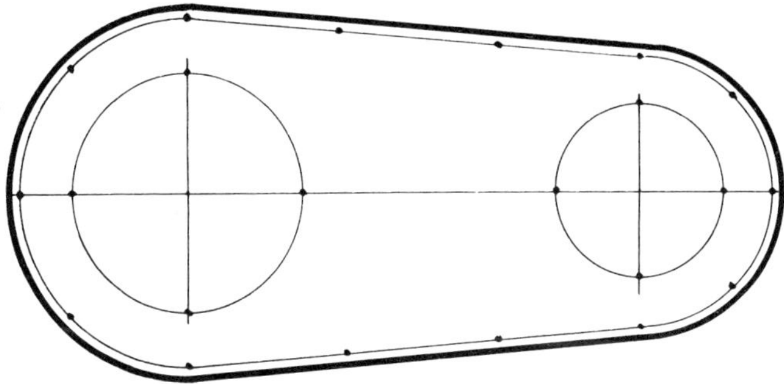

Fig. 32

Fig. 33

Automatic Centre Punches

Of late years a form of centre punch has been developed that dispenses with the use of a hammer. This is the so-called automatic centre punch illustrated in *Fig. 34*. Two forms are depicted. That at A is operated by pressing down on the point and continuing the pressure until a hammer device, inside the barrel of the punch, is automatically released to strike the base of the punch and so make the punch mark required. The hammer device is powered by a spring, also inside the barrel, and this spring may be adjusted by turning the cap of the tool in order to strengthen or weaken the spring pressure.

The punch seen at B is a much simpler affair. It, too, is spring oper-ated; but only by the workman pulling up the cap to which a spring inside the barrel is attached and then letting it go. The punch marks used for outline work or for defining centre lines need only to be of a light character; on the other hand those used for denoting the centres at which holes are to be drilled need to be heavier in order that the drill shall start truly. In this connection it should be added that unless the hole required is small it must first be pilot-drilled. That is to say a small

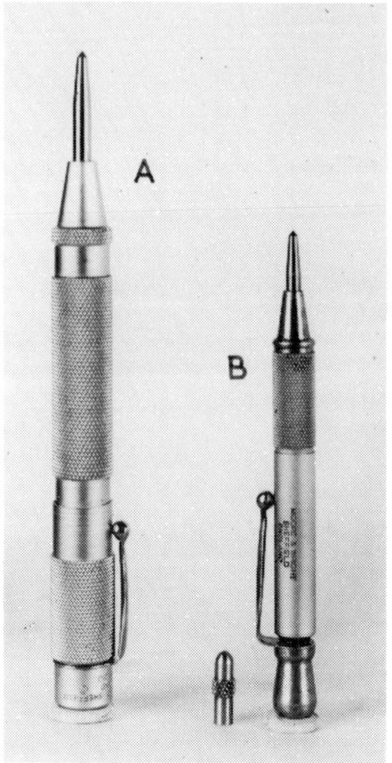

Fig. 34

drill is first used and this is followed by one of a size required to produce the hole desired. In this way chances of the hole running off-centre are largely minimized.

Mention must also be made of the combination centre drill now largely used for starting holes in work. This tool, depicted in *Fig. 35* is intended for use in machines only. Its construction is such that its employment in hand drills would only lead to breakage.

As will be seen the drill is double-ended and comprises a pilot drill of short length grafted on a shank that is itself a drill.

These drills are made in several sizes, at the moment of writing in both inch-fractional and metric dimensions, commencing with a size bearing an ⅛ in. body and 3/64 in. pilots. These dimensions should lend point to the previous remark that the combination centre drill is for use

Fig. 35

in machines only. In all cases the countersink portion of the drill is ground to an included angle of 60 degrees to accommodate a standard lathe centre.

The Surface Plate

A number of operations in the metal working shop require the use of a reference surface that is perfectly flat. For example, if the worker needs

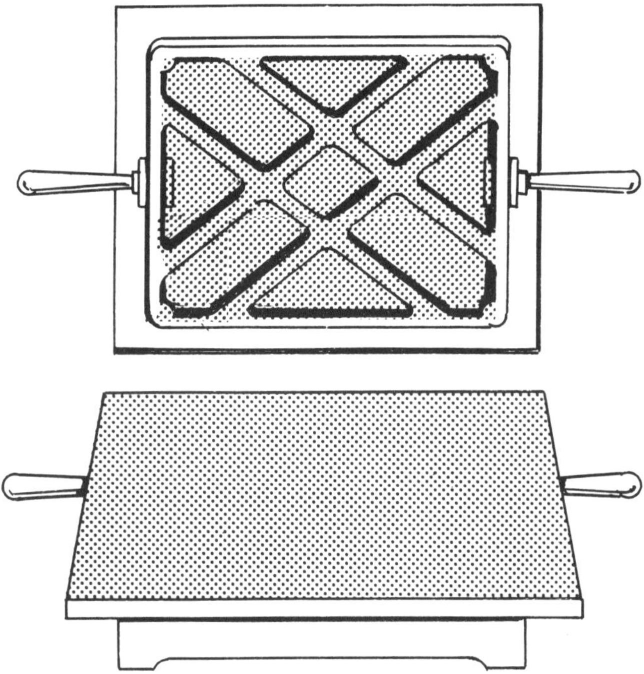

Fig. 36

to make sure that the face of some component on which he is working is perfectly flat, he can place it on the reference surface for this purpose. If, having done so, there is in the first instance no 'rock' observable, he can then proceed to ascertain the extent of the area of contact between the reference surface and the component itself. To do so the reference surface is very thinly smeared with 'marking blue' which will adhere to the face of the component when it is moved about the reference surface. The extent of the blue marking adhering to the work will indicate to the mechanic the extent of the contact area. If this is to be increased the mechanic applies a scraper to the high spots indicated by the marking, retesting the work from time to time and continuing the scraping operation until the required area of contact has been obtained.

The reference surface mentioned is called a surface plate. In essentials it is an iron casting, ribbed on its underside in order to obviate distortion, and machined on its top side to present a flat surface.

The size varies. Some measure only a few inches square, others are veritable tables at which several men may work. A representative surface plate is illustrated in *Fig. 36*.

Fig. 37

It will be seen that a pair of handles are attached to this surface plate. Not only is their purpose to help in lifting the surface plate, but they are sometimes useful when taking the plate to work that needs testing, but may be too large to bring to the surface plate itself.

In many shops, however, the most important use for the surface plate is when marking out work set out upon it. A typical arrangement is depicted in the illustration *Fig. 37*. For those who do not wish to go to the expense of a cast iron surface plate an adequate substitute, for marking out purposes only, is a piece of plate glass. This should be at least ⅜ in. thick and should have its edges and corners dressed by grinding on a sandstone wheel to get rid of any sharp surfaces. Additionally the plate can, with advantage, be set in a frame formed from a piece of plywood, as a base, with its sides made from strips of hardwood.

The Scratch Gauge

Those who have had some grounding in woodwork will be acquainted with the carpenter's scratch gauge used for marking tenons and the like. Faced with some work on sheet metal, where the use of a scratch gauge would have been an advantage, the author made up from scrap the device depicted in *Fig. 38*.

Fig. 38

Its construction should be apparent from the illustration where it will be seen that the extension bar that passes through the metal fence carries a gramophone needle to carry out the marking-off on the work. As the old '78' gramophone needles are now so difficult to obtain a compass point may be used instead.

CHAPTER II

HOLDING DEVICES

AN ESSENTIAL PREREQUISITE to good workmanship is the ability to hold the job securely whatever operation is to be performed upon it. As, for the most part, we are concerned here with hand work it is the bench vice and other ancillary holding devices that we shall be considering.

The leg vice illustrated in *Fig. 1* is in all probability the forerunner of the engineers' vices in use to-day. It was furnished with a soleplate for bolting to the bench and a leg reaching down to the floor as a support. Contrary to modern practice the jaws do not remain parallel at all times for, as will be seen, the moving jaw is hinged and so describes an arc. Not many examples of the leg vice are likely to be encountered today.

Engineers' Vices

Bench vices come in many types and sizes varying from the light table vice illustrated in *Fig. 2,* to the quick-acting vice which will be described later. As its name implies the table vice is intended for temporary attachment to an ordinary table so, to that end, it is only provided with a hand clamp screw and pad to hold it in place. Naturally, so light a vice cannot be expected to cope with heavy work; but it is well worth a place in any tool kit.

The vice illustrated in *Fig. 3* is an enlarged version of that just described. In this case, however, the vice is designed for permanent mounting on the bench and is fitted with a base that allows the body of the vice to be set round at an angle when required.

The body and the moving jaw castings are so designed that they will grip round material when required.

The vice seen in the illustration *Fig. 4* is the standard vice to be found in all engineers' workshops. It is, of course, designed for permanent bench mounting. In company with all but the lightest of vices the jaw inserts, which are hardened, are detachable and are serrated in order to afford a grip. The standard vice has no quick-acting opening or closing mechanism. On the other hand the vice depicted in *Fig. 5* has such a

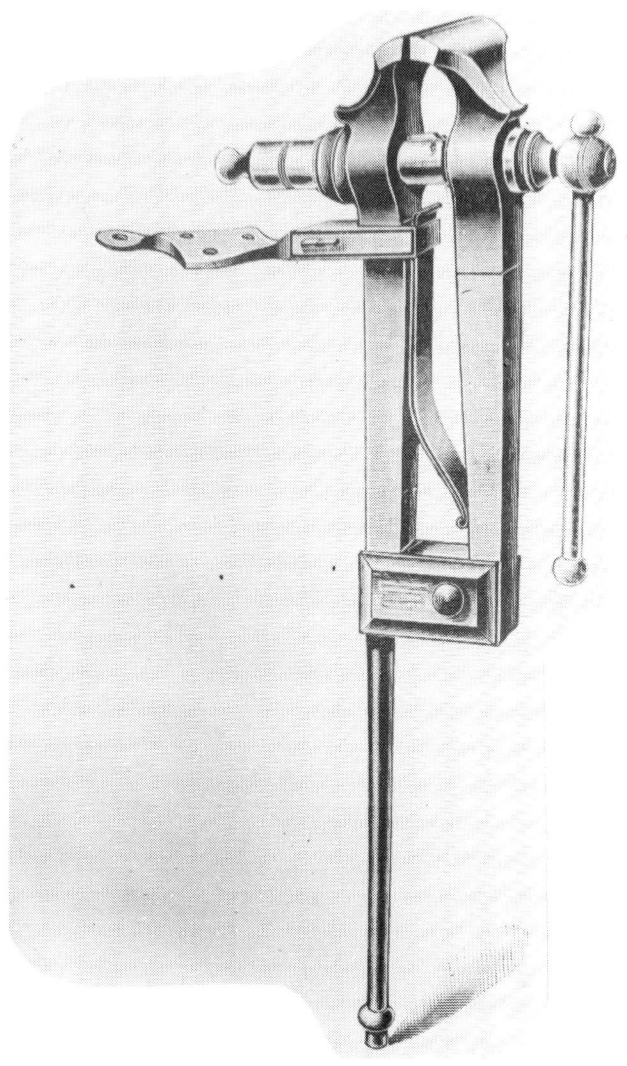

Fig. 1

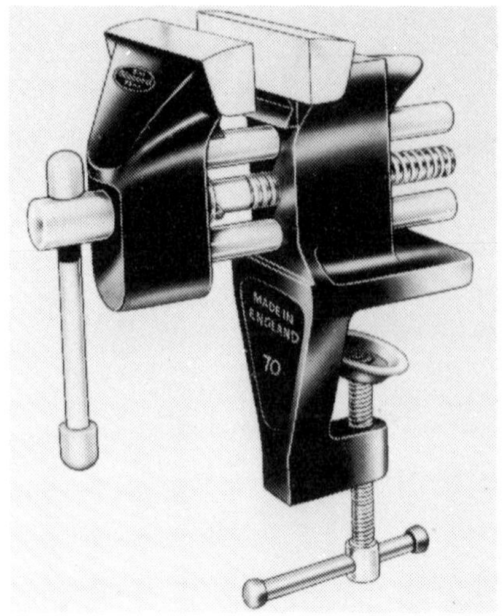

Fig. 2

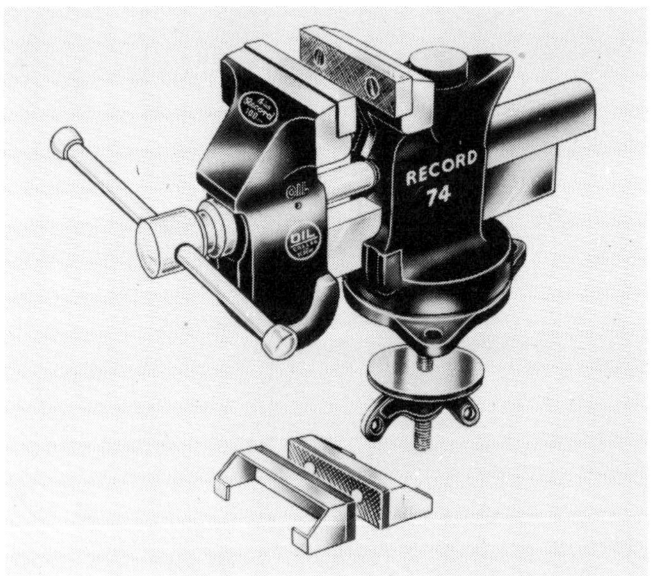

Fig. 3

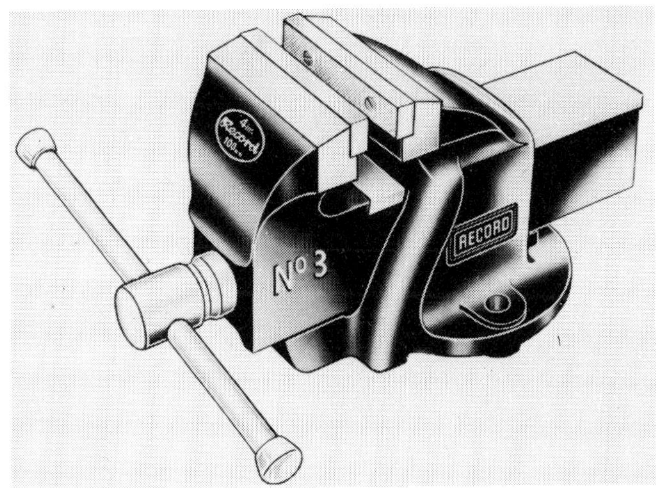

Fig. 4

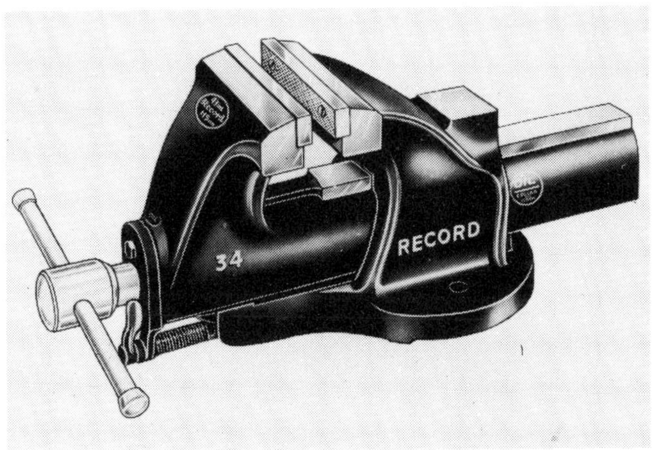

Fig. 5

provision. The mechanism is operated by the short lever seen just below the hub of the actuating screw on the left of the illustration.

Whereas, in the standard vice, the nut engaging the screw is fixed, in the quick-acting vice the nut is moveable and may be disengaged, thus releasing the free jaw so that the vice may be adjusted approximately before securing the work in the ordinary way.

The mounting of the vice on the workshop bench is a matter of some importance if comfort in working is to result.

The diagram, *Fig. 6*, demonstrates the two cardinal points to be observed. As to the position of the vice on the bench top, it is best set

Fig. 6

directly over one of the legs to ensure that the maximum rigidity is obtained.

Instrument Vices

Much of the work carried out in some engineering workshops is of too light a nature to be entrusted to a large and heavy vice. For this reason special vices, known as instrument vices, have been introduced to accommodate this class of work.

The first example is the 'Eclipse' instrument vice seen in the illustration *Fig. 7*.

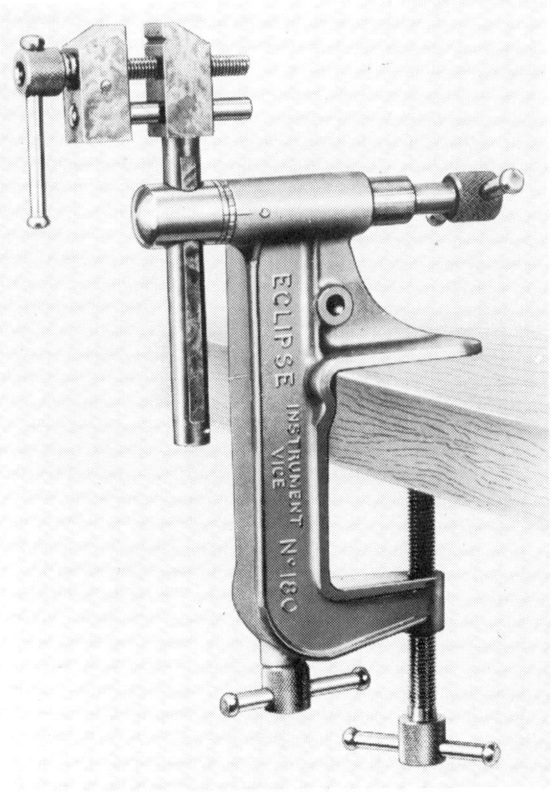

Fig. 7

This fitment, as will be observed, is for table mounting or for attaching to a short piece of wood that may be caught in the main bench vice if thought convenient. The illustration *Fig. 8* will give some idea of the versatility of this particular tool.

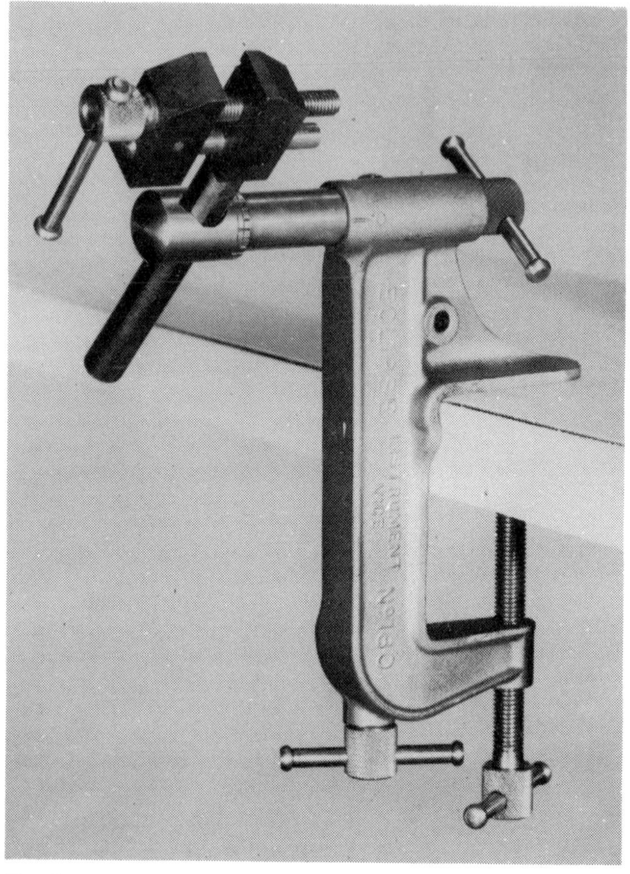

Fig. 8

A somewhat heavier instrument vice is illustrated in *Fig. 9*. This was made by the Offen Company who have now, alas, gone out of production. The base of the device may be turned through 350 degrees and

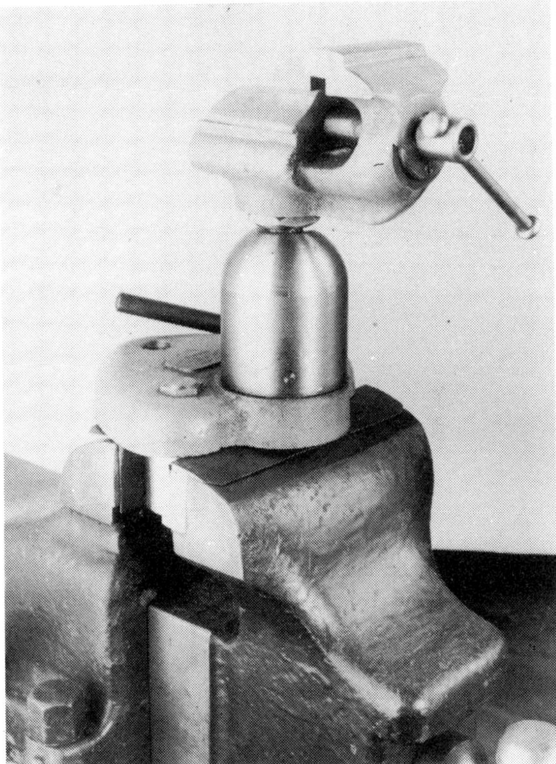

Fig. 9

locked at any desired point whilst the vice as a whole, which is mounted on a large ball, can be swung through 90 degrees so that the edge of the vice remains uppermost if needed.

Vice Clams

We have already noted that the jaws of many vices are provided with inserts that have serrated surfaces. These are intended to provide a firm grip on the work. However, if it is finished or semi-finished work that is set in the vice, some protection against the roughened surface is essential. According to class of work in progress these vice clams as they are

Fig. 10

Fig. 11

called may consist of pieces of card, bent at right angles so that they may stay in place on the vice jaws, or they may take the form of copper or lead sheet bent up at right angles as seen in the illustration *Fig. 10*.

Of late years detachable red fibre liners for vice jaws have been introduced. These take the form depicted in *Fig. 11* and *Fig. 12*. They consist of light steel formers, bent up as shown, having red fibre facings attached to them by light aluminium rivets. These liners wear well, as the very old pair in the illustration will perhaps confirm. They are obtainable in sizes to suit vices from 3 m. to 6 m. wide.

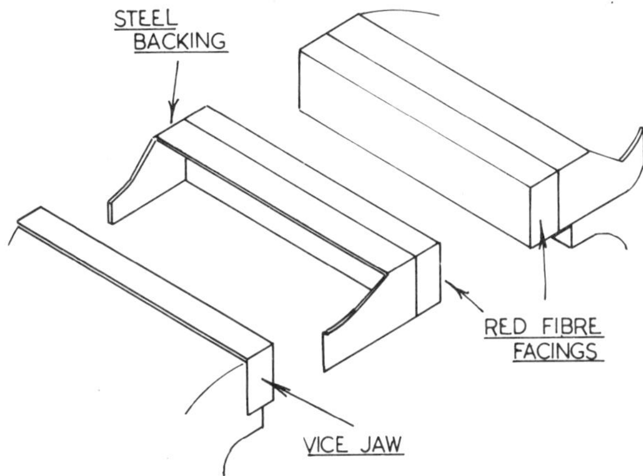

Fig. 12

Toolmakers' Clamps

It is sometimes necessary to clamp two parts together so that they may
be drilled. It is then that the toolmakers' clamps seen in *Fig. 13* come
into play. They are obtainable in several sizes, two of which are illustrated. These clamps are similar to those known to the woodworker.

Fig. 13

G-Clamps

If large components need to be held together temporarily, the worker
makes use of G-Clamps illustrated in *Fig. 14.* These vary widely in

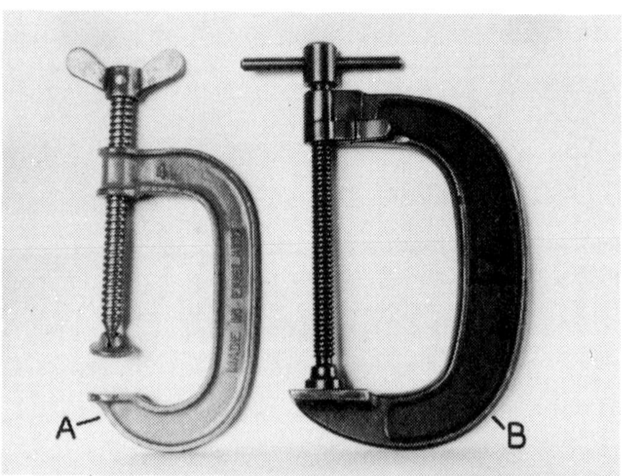

Fig. 14

capacity. The clamp depicted at A is fixed and can only be opened or closed on the work by turning the wing nut set at the top of the screw. On the other hand, the clamp seen at B is provided with quick-release mechanism operated by the trigger at the top of the bow. The device enables the user to adjust the clamp quickly before setting it to grip the work.

Fig. 15

Both clamps are fitted with an 'elephant's foot' that helps to align the tool when in use.

The Woodworker's Vice

Finally, a vice that is not strictly intended for the metal worker. This is the Record Junior Vice illustrated in *Fig. 15* made for the use of the woodworker. In this context the fitment needs to have wood linings as with any other woodworkers' vice. But without them, and mounted in the manner seen in the illustration, the vice can sometimes be of service to the metal worker.

Casting Vice Clams

To revert for a moment to the lead vice clams already mentioned. For those who needed something rather heavier than can be made from sheet lead the device illustrated in *Fig. 16* was once available, and indeed may well still be so. The mould is hinged whilst the ladle has a duct leading directly into the mould. Therefore, to cast a vice clam, after locking the two halves of the mould together, one need only melt sufficient lead in the ladle then tip the device to let the molten metal run into the mould. Once the assembly has cooled sufficiently the mould can be opened and the clam tipped out.

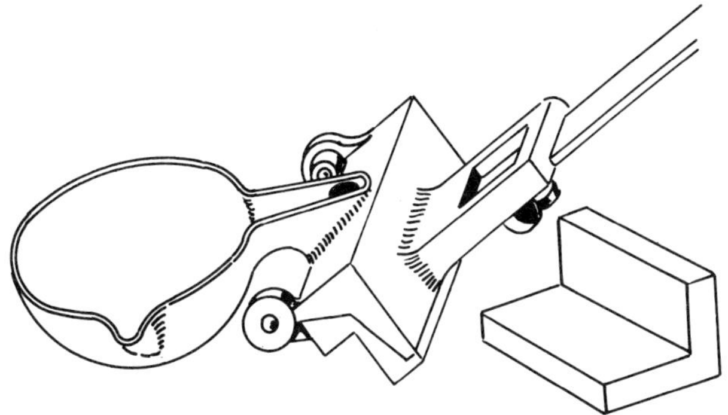

Fig. 16

In order to achieve a somewhat comparable result the author, many years ago, devised the simple fixture depicted in *Fig. 17*. The illustration should explain itself. As will be inferred from the diagram, casting is carried out with the fixture set in the vice.

Bright mild-steel
3/8"square rivetted to angle
2" x 1/4"
Angle iron
Plate 4"x 2.5/16"x 3/16"
Dowells

Fig. 17

SPRING
GRUB
SCREW
BRASS PAD
SCREW

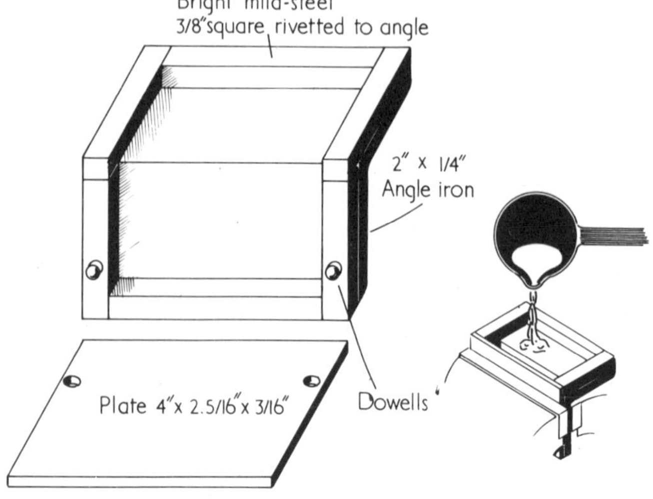

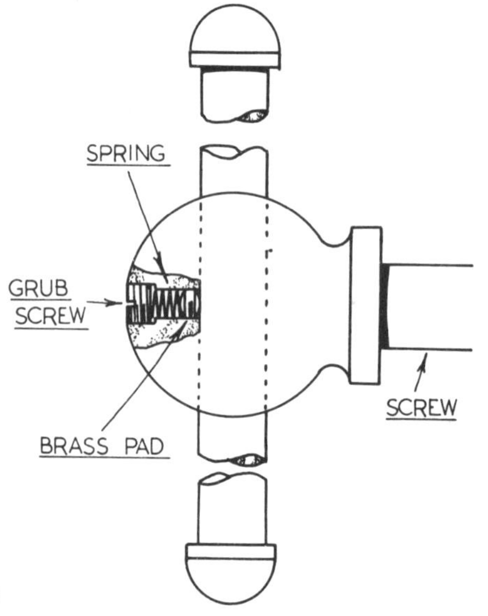

Fig. 18

A small modification to the screw of any vice may be found advantageous. This addition is illustrated in *Fig. 18*. The screw is removed from the vice temporarily and is drilled and tapped to take the spring and pad used to apply frictional pressure to the handle in order to secure it; so preventing the upsets at the ends of the handle from pinching the fingers. Only light pressure is needed to restrain the handle which does not, of course, need to be locked.

Before closing any remarks on the subject of the vice and the methods of holding work a word should be said about holding thin work. It will be clear that such parts cannot be gripped directly between the jaws of the vice without fear of distorting them. Instead, the work should be set on a piece of wood, itself held in the vice, and secured by panel pins against movement as depicted in the illustration *Fig. 19*.

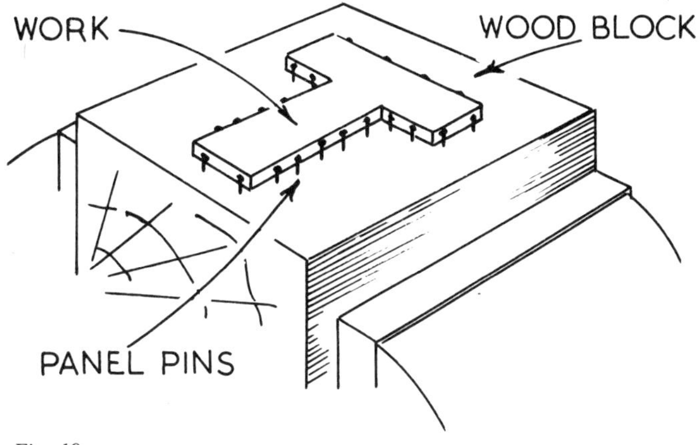

Fig. 19

CHAPTER III

CUTTING TOOLS

Chisels

ROUGH SURFACES OR projections on castings often need to be removed before work upon them can be resumed, and it is for this purpose that chipping, or 'cold chisels' as they are usually called, are employed.

The chisels, for the most part, are made from cast steel, hardened and tempered to withstand the hammer blows they sustain. Modern steel-making practice however has produced a material that is far superior as regards wear and can be sharpened by filing.

Cold chisels take several forms: those used for the straightforward chipping away of projections or rough surfaces are shaped as shown in

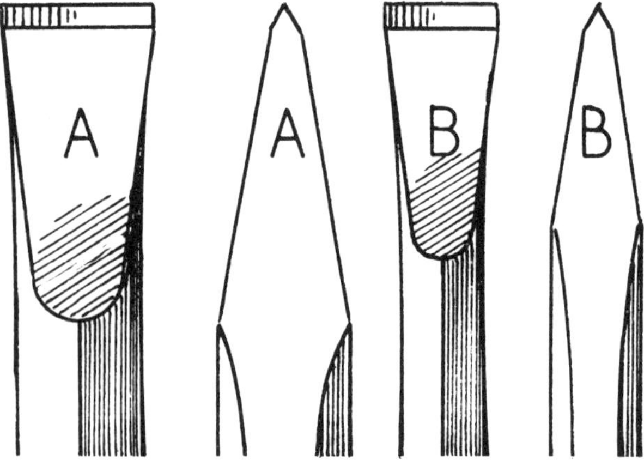

Fig. 1

Fig. 1 at A and B. The latter is termed a 'cross-cut' chisel and is used for cutting a series of channels or grooves over a wide surface so as to ease the work of the flat chipping chisel which is used subsequently to remove the metal between the grooves. Nowadays the necessity for a complete set of good quality cold chisels is not by any means as great as it used to be. The advent of electrical power, and the light machine tools to use it, are factors which have largely made the cold chisel redundant.

Nevertheless there are occasions when the use of the chisel is justified, and indeed may overcome some difficulty. For this reason, then, note should be taken of the composition of a full set of cold chisels and the work they are able to perform.

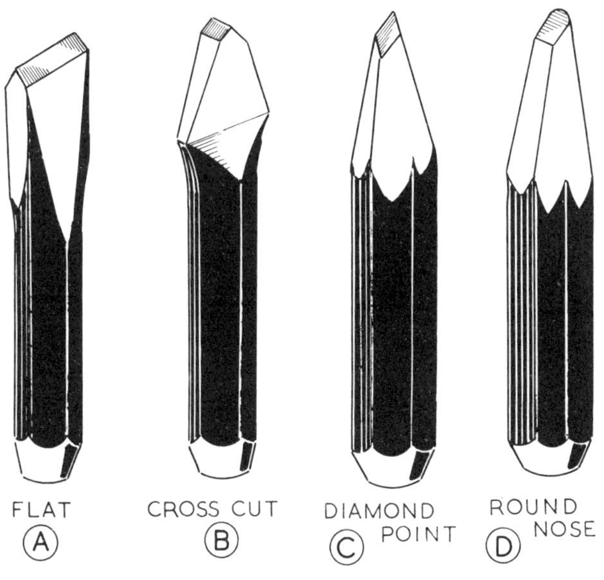

Fig. 2

Chisels A and B in *Fig. 2*, we have already considered for they are the flat and the cross-cut pattern respectively. The chisel depicted at C has a diamond-shaped point and is useful when clearing out corners. The pattern D, a half-round chisel, is the least often used. The relative importance of the several tools is in accordance with the order in which they appear in the illustration.

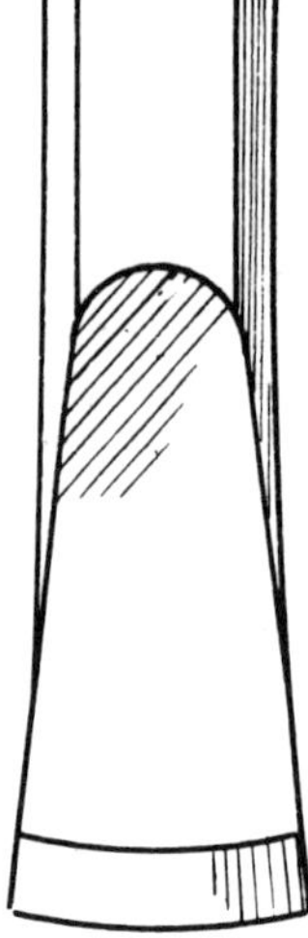

Fig. 3

As to the flat cold chisel A, a final word must be said before the subject is left. Although the cutting edge is represented as a straight line in the illustration, in reality the tool will cut better if the edge is very slightly rounded as seen in *Fig. 3*.

Whilst it is, of course, of the utmost importance to ensure that the business end of a cold chisel is in good condition, it is also essential that the opposite end of the tool is also clean and free from 'rags'. These result from continual hammer blows which tend to upset the end of the chisel and cause rags to form. If not checked they can become dangerous and damaging to the hand when detached by a blow from the hammer. *See Fig. 4.*

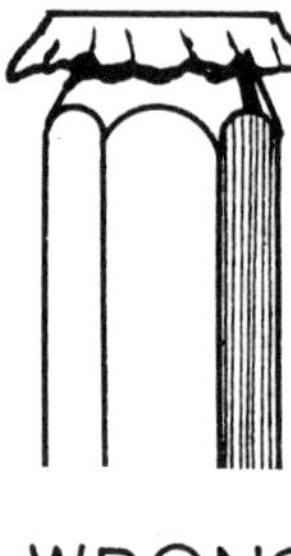
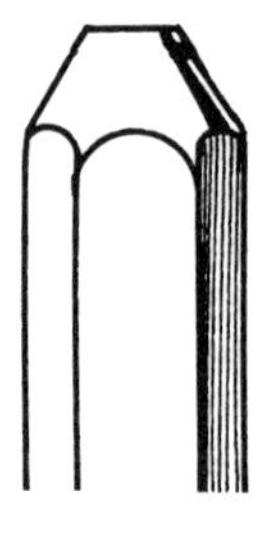

Fig. 4

Files

Files form one of the most important class of tools in the equipment of
the small or amateur workshop. It follows, therefore, that their selec-
tion must be made with care or the worker may find himself with a
battery of files some of which, for one reason or another, are unsuited
to his requirements.

Files are classified according to their shape and also according to the
coarseness or otherwise of their cutting teeth. In the first classification,
shape; the following are the principal forms: flat, half-round, round,
square, triangular or 3-square. There are also files for specialized work
such as key-cutting, but these can be considered when the need arises.

The accompanying table shows the uses to which the various shapes
of file can be put:

Shape	Use
flat	for filing up and finishing flat surfaces
half-round	for working on the inside of curved surfaces and large holes
round	for enlarging holes and slots
square	for filing out keyways, slots and enlarging square holes
triangular	for cleaning out the corners of slots, square holes, and keyways, also for sharpening saws

As has already been said, files are further classified according to the
coarseness of their teeth. These classifications are:

1. rough; 2. bastard; 3. second-cut; 4. smooth; 5. dead smooth.

The worker must therefore choose carefully both the shape and
coarseness of the files he is likely to use. Here it should be noted that
files for use on brass and bronze should be kept separate from those
used for cast-iron and steel, remembering that a file which has become
too blunt to cut brass will still be serviceable for filing cast-iron and mild
steel.

For dealing with aluminium alloys a special file has been developed.
These materials, if cut with ordinary files, tend to 'pin'; that is the teeth

of the file tend to fill up with metal which is hard enough to damage the surface of the work. To some extent the trouble may sometimes be minimized by 'chalking' the file, using a lump of natural chalk or a piece of blackboard chalk for the purpose. The chalk is rubbed on the surface of the file, filling up the teeth and preventing the aluminium filings adhering to them.

The special files referred to, sometimes called milling files, have teeth set radially across them and are cut in the form depicted by *Fig. 5*. Files of this form cut freely and seldom 'pin'. However, some experiments carried out by the author, which deposited hard chromium on the file, seem to have disposed of even this minor difficulty.

Leaving aside any specialized requirement, what advice can be given to a newcomer to help him select a practical collection of files that will cover the class of work he is most likely to encounter?

The worker in the small shop is seldom likely to need a rough file so the list that follows is based on this omission.

Length in inches	Type
12	one flat bastard file
12	one second-cut file
10	one flat bastard file
10	one second-cut file
8	one flat bastard file
8	one second-cut file
8	one flat smooth file
8	one half-round bastard file
8	one half-round second-cut file
8	one round file
8	one three-corner file
8	one square file
6	one square file

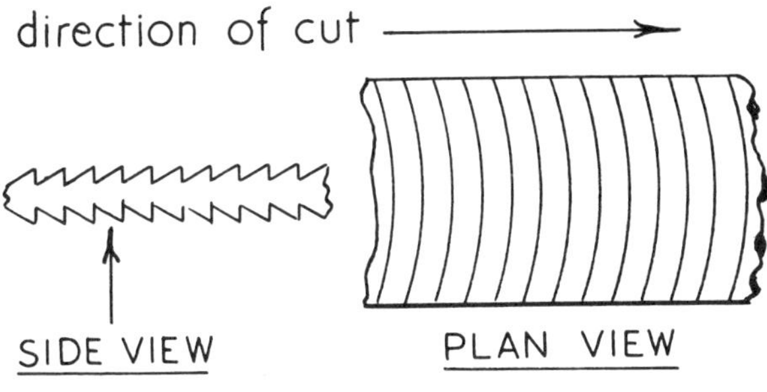

Fig. 5

In addition a small number of 4-inch and 5-inch files of varying types will be needed to deal with the small components the worker is likely to encounter.

Storing Files

Files should at all times be stored in racks and not jumbled together in a drawer. The author prefers to house large files horizontally whilst the small files are set vertically in racks specially made for the purpose.

Many flat files are made with a 'safe' edge. The term should explain itself. It means that one edge of the file is devoid of teeth enabling the user to cut into a corner without damaging any finished surface.

File Handles

Proper file handles must at all times be fitted to any files in use. Without them there is the risk of serious injury if perchance the hands come into

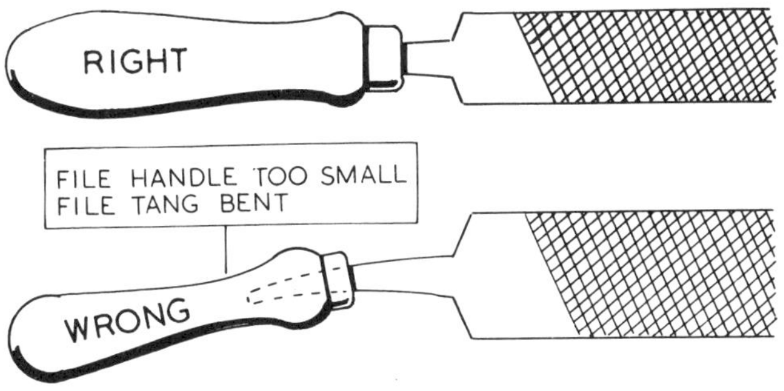

Fig. 6

contact with an unguarded file tang. It is essential that the correct size of handle is fitted to a file and that it is put on straight. The correct and incorrect alignments are illustrated in *Fig. 6*.

A further word about filing. Never use a new file on the surface of a casting; the skin is generally hard so an old file should be employed to prepare the way for the new one. Severe cases of hardness may need treatment with a hammer and cold chisel before filing can commence.

Avoid allowing the work surface to come into contact with oil or grease which would slow down the filing operation. The same remarks apply to the hands upon which there is usually sufficient residual grease to produce the same result.

File Card

As to the files themselves they need brushing from time to time in order that any encrustations or particles of metal can be cleared away from their surface. The brush to be used is made from a strip of file card usually purchasable from tool merchants. File card is used in the carding engines of the wool industry, hence its name. It consists of a steel wire complex set in a fabric matrix, the wires projecting from this base. The material can readily be tacked or glued to a flat wooden handle to form a wire brush. See *Fig. 7*.

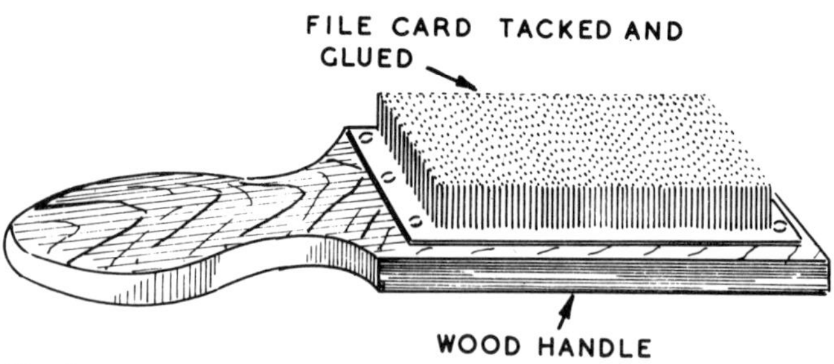

Fig. 7

Methods of Filing

Finally, always make sure that the work is held firmly, and is gripped in the vice in such a way that no vibration takes place during the filing process. The noise vibration produces is unpleasant and the vibration itself retards progress.

Two methods of filing are illustrated in *Figs. 8 and 9* respectively. In the first, the file is shown passing diagonally across the work in one

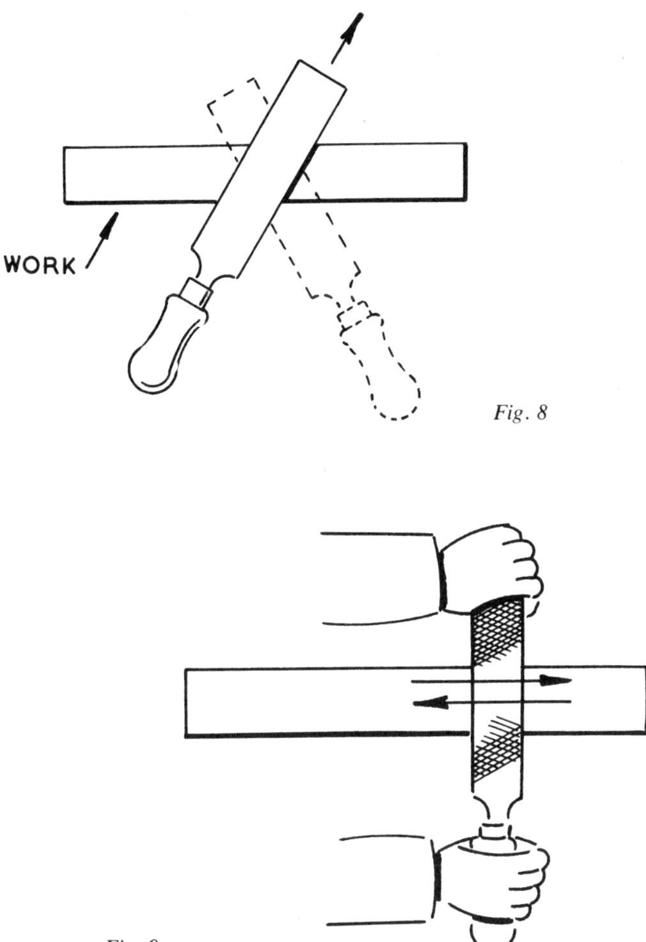

direction and then being applied again diagonally, in the opposite direc-tion. This is a practice useful when heavy filing work is undertaken; the crossing of the file strokes tending to cancel out any pattern that would otherwise be formed.

The illustration *Fig. 9* represents the action of 'draw-filing'. A useful technique when finishing small flat surfaces or the edges of components. Care must be taken, however, to ensure that the file is not canted during the operation or the work may become rounded.

Saws

In metal working two basic saws only are needed. Of these the most important is the hacksaw which is used for cutting material to length, rough-cutting work to shape, or in fact for all purposes in which sawing is involved.

The hacksaw comprises a metal frame provided with a handle and means of securing and tensioning the saw blade. At one time hacksaw frames were fixed and would only accept the length of blade for which they were intended; today frames are made adjustable in order to take the small range of blade length now available. A typical adjustable hacksaw frame is illustrated in *Fig. 10*.

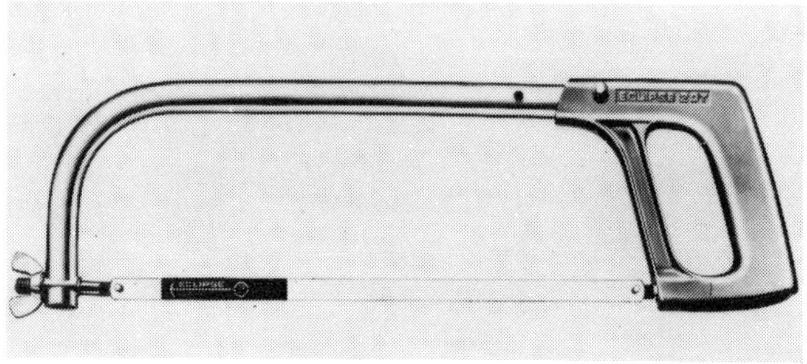

Fig. 10

Modern production has limited the length of blade for use in the hand hacksaw to 10 inches and 12 inches with tooth-pitching varying from 14 to 32 teeth per inch. The blades are made from both high-speed and alloy steel, and there are also available flexible blades that are less likely to break when in the hands of less experienced workers.

The diagram *Fig. 11* shows the characteristics of the standard hacksaw blade, and the location of the points from which dimensions given in the makers' lists are taken.

In this connection it is perhaps worth noting that current lists now give these dimensions in both inches and millimetres, and that the length of the saw blade is taken over the outside of the pin holes used for mounting the saw and not from the hole centres as might possibly be imagined.

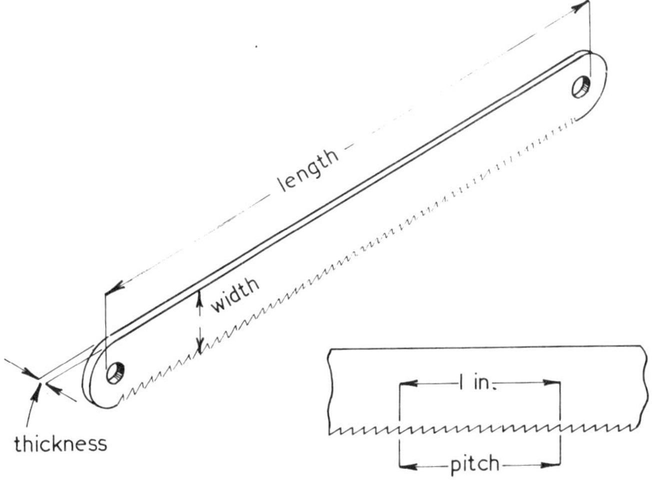

Fig. 11

The tooth-pitching has been standardized according to the accompanying table where the use of each pitch is given:

Pitch	Use
14	a course pitch giving extra clearance for mild steel
18	a medium pitch for use on hard steels; less chip clearance
24	a fine pitch for use on small sections
32	an extra fine pitch for use on the smaller sections

For some years now James Neill of Sheffield have marketed a small saw they call the 'Eclipse' Junior Hacksaw. The frame of this saw accommodates blades 6 in. long by ¼ in. wide, having a pitch of 32 teeth to the inch. These saw blades have a wavy set to ensure adequate clearance and are made from a flexible steel that assures the user a satisfactory endurance factor. Over the years that the Junior Hacksaw has been in being minor developments and modifications have been made. Now it somewhat resembles the carpenter's tenon saw and it has all the stability of that saw. It is depicted in the illustration *Fig. 12.*

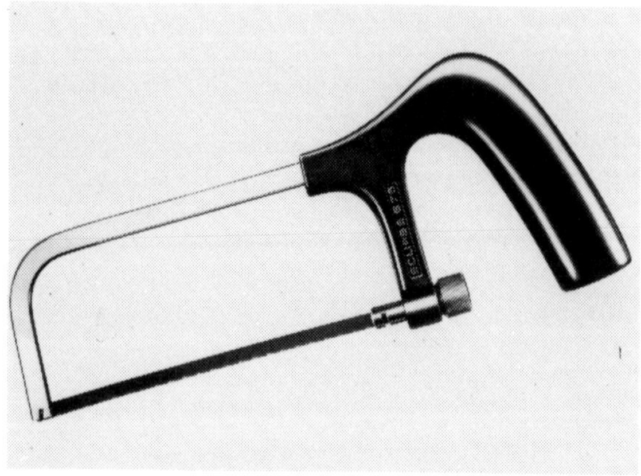

Fig. 12

The Piercing Saw

The metal worker is sometimes called on to cut intricate patterns in sheet metal or metal plate. The work being analogous to fretsawing in wood. The saw used is the piercing saw illustrated in *Fig. 13*. It comprises an adjustable frame in which special blades can be set under tension. The blades themselves vary widely both as to width and tooth-pitching; so the manufacturers have provided a comprehensive list, in some cases an actual collection of examples, that will give the potential user a basis for selecting the correct blade for the work in hand.

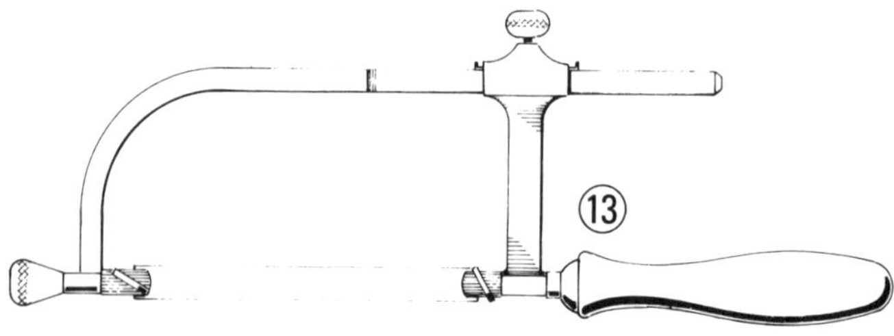

Fig. 13

As will be noticed the saw blades are mounted in holders provided with thumb screws. These put tension on metal plates forming part of the frame and thus on the saw blades themselves.

Shears

For cutting thin sheet metal shears are used. These resemble a pair of household scissors but the leverage between the handles and the blades is far greater, as befits the work shears are called on to perform.

For the most part the worker needs to make cuts in straight lines only, and for this duty the pair of shears illustrated in *Fig. 14* is used. As will be seen the blades are straight so the cutting of internal curves is not possible, though, with a little juggling, one may deal with external ones.

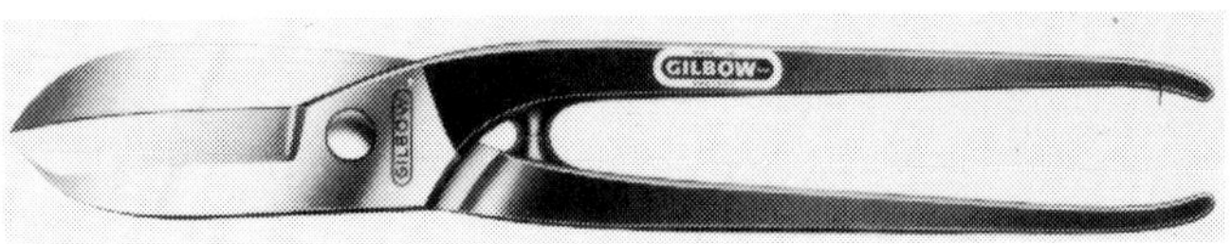

Fig. 14

However, there are pairs of shears specifically designed for cutting curves, both internal and external.

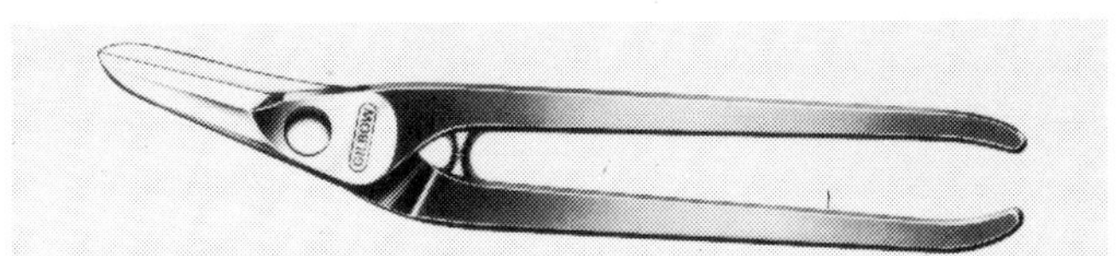

Fig. 15

These shears are depicted in the illustration *Fig. 15*. Some sheet metal is too thick or too resistant to shearing to be handled by the type of shears just described. For this work the bench shears illustrated in *Fig. 16* are used. As will be appreciated the leverage available is massive when compared with that of a pair of hand shears. The bench shears seen in the illustration are suitable for straight cutting only, though one may of course 'nibble' around the circumference of external curved surfaces.

If the bench shears are only to be used infrequently it is convenient, as may be observed from the illustrations, to provide a mount so that they may be set in a strong bench vice.

Also, when the material to be cut is of greater length than is usually convenient, one may extemporize an extension below the lower shear

Fig. 16

Fig. 17

blade by way of support. As may be seen in *Fig. 17* this comprises a length of mild steel set end-on and secured by a pair of the toolmaker's clamps illustrated in the previous chapter.

CHAPTER IV

EQUIPMENT FOR DRILLING

Drilling and Reaming

THE DRILLING OF holes is one of the more important, and certainly one of the most frequent operations that the metal worker undertakes.

At one time the diamond-pointed drill was the only tool available. It had some sort of advantage in that it could be made by the worker himself, but its accuracy was low and its cutting speed the same.

As a matter of interest it is depicted in the illustration *Fig. 1*. The drill can be made from silver steel; knocked up into shape by forging, then filed to form and finally hardened and tempered.

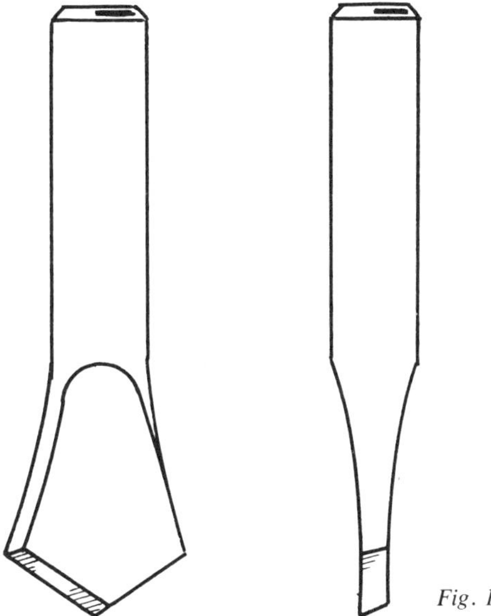

Fig. 1

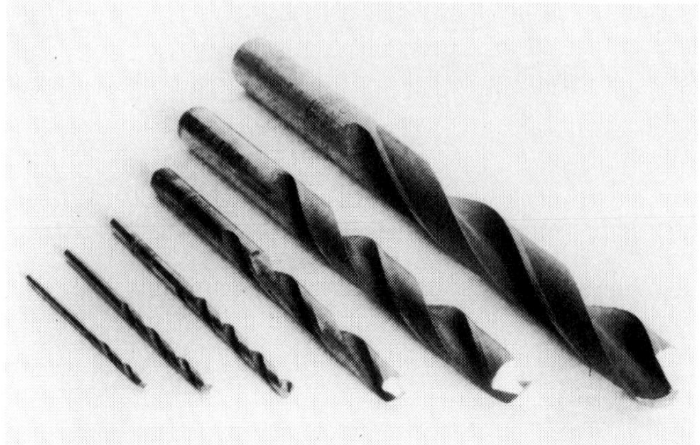

Fig. 2

Whilst this is a type of drill that must now be considered archaic, it should be noted in case it might be used to overcome a workshop difficulty.

The type of drill now commonly employed is the twist drill illustrated in *Fig. 2* and in *Fig. 3* at A.

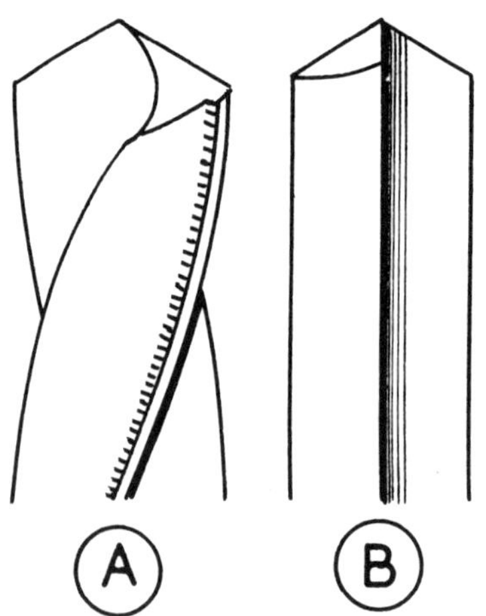

Fig. 3

Twist drills are obtainable that have a variety of spiral formations. These are depicted in the diagram *Fig. 4*. Drills are usually supplied

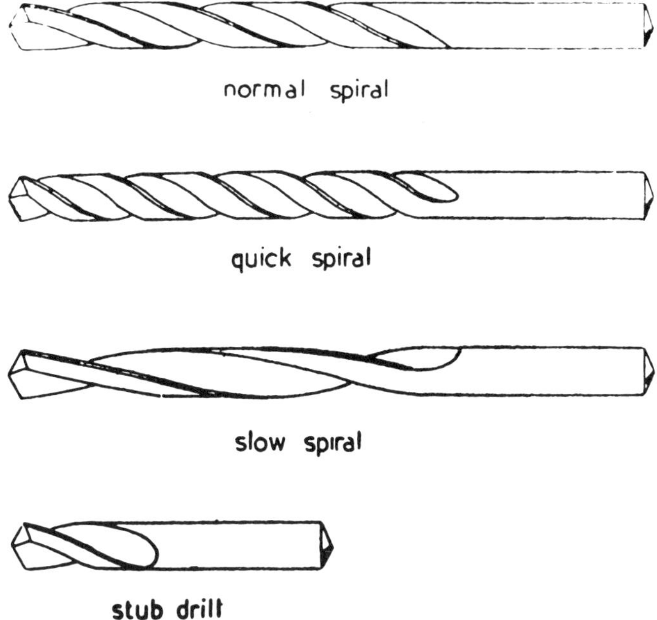

Fig. 4

having the normal spiral seen at the top of the illustration. Quick spiral drills are used on aluminium alloys whilst the slow spiral drill has largely replaced the straight-fluted to be described later. The stub drill depicted at the foot of the diagram is useful on machines having restricted space, or in conditions where a stubby and rigid drill is essential. The standard length of drill found in the toolshops is classed under the heading 'jobbers', but long series drills are also obtainable. These vary widely in length.

The cutting edge of the twist drill, as the result of the helical nature of its construction, has plenty of top rake.

This means that the body of the drill behind the cutting edge is raked at an angle as depicted in *Fig. 5* at A. It is this condition that makes the twist drill eminently suitable for drilling ferrous metals and aluminium alloys, but not for drilling the bronzes and brass in which the twist drill tends to grab.

On the other hand the drill illustrated in *Fig. 3* at B is definitely suitable for the drilling of brass. This is the straight-flute drill once

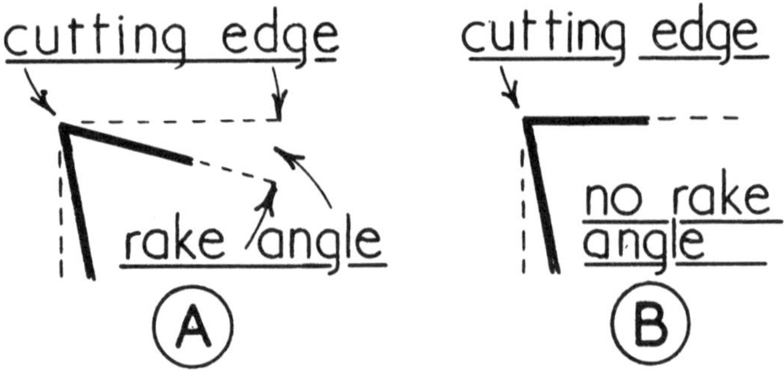

Fig. 5

largely used by brass workers. The body of the drill behind the cutting edge subtends no rake with that edge; consequently the drill does not grab when working in brass.

It is said that the late Sir Henry Royce, who had started his commercial career as a manufacturer of electrical equipment, would countenance no other form of drill in his brass-finishing shop at Rolls-Royce, because of his dislike of the twist drill for brass-working.

Be that as it may, the straight-flute drill has largely disappeared in favour of the universal employment of the twist drill.

In any case a small modification to the drill point fits the tool for handling brass. For this reason those workers who regularly drill the material tend to keep a set of modified drills for the purpose.

Modifying the Drill Point

The modification referred to simply consists in stoning or grinding the face of the drill below the cutting edge in order to reduce the top rake to nil. The position is defined in the diagram *Fig. 6*.

As to availability: nowadays twist drills may be obtained in sizes varying from a few thousandths of an inch to several inches in diameter. These are grouped in classes termed Inch Fractional, Letter, Number and Metric; and it is this last group that, as a result of metrication, is likely to supersede all others.

The double-ended drill illustrated in *Fig. 7* is designed to enable the user, when employing a drilling machine or lathe, to start a drill hole before opening it out to the required size. It is also intended to drill seatings for lathe centres upon which work is to be turned. It is *not* intended for use in hand drills. These devices originally produced by the Slocomb Company of America are usually referred to as 'centre drills'.

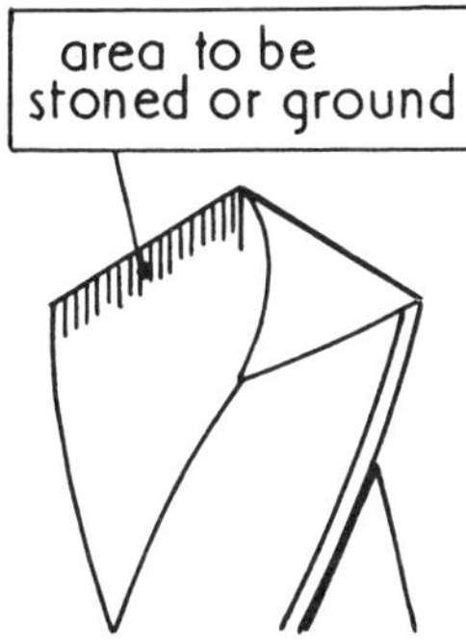

Fig. 6

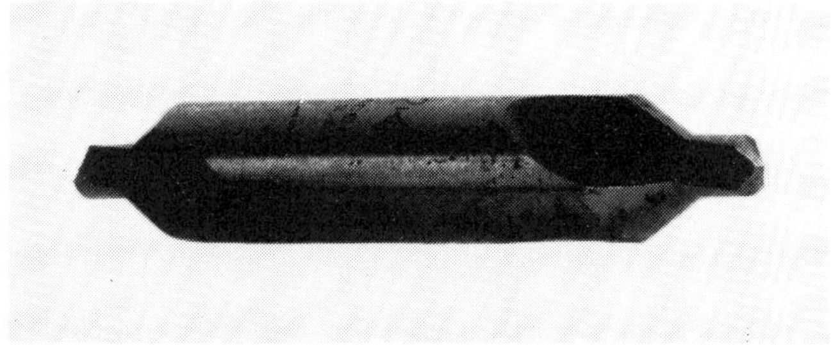

Fig. 7

Tipped Drills

Before leaving the subject of twist drills in general, mention must be made that some drills have their cutting edges tipped with tungsten carbide. This is a material of great hardness but little mechanical strength. Tools tipped with tungsten carbide are capable of cutting hard substances such as glass, some stonework and the very high tensile steels; consequently tipped drills are in use by building workers as well as in the metal workshop.

Countersinks

Workers are often required to modify a drilled hole in order to fit it so that it will take a countersunk screw. The cutter used for this purpose is depicted in *Fig. 8*. For the most part the included angle of the point is 90 degrees and there are four cutting lips. Cutters of this type, if run too fast, are apt to 'chatter' and so produce an uneven surface to the work.

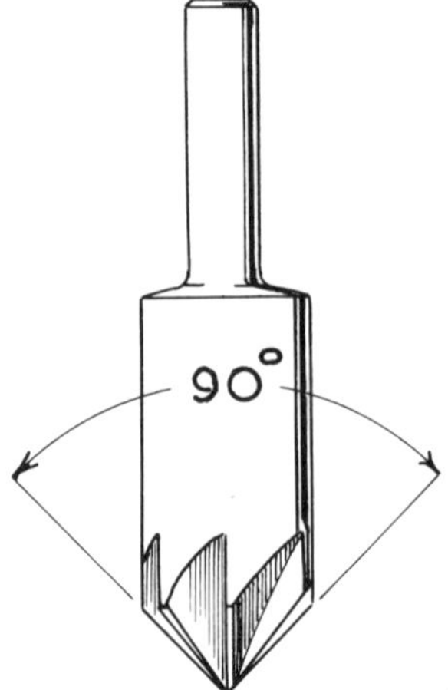

Fig. 8

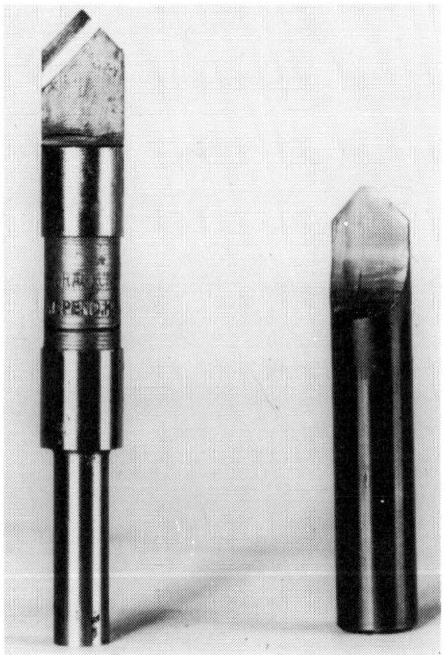

Fig. 9

To overcome this difficulty the device depicted in *Fig 9* has been introduced. This countersink has but one cutting edge formed by machining a flat surface on the body of the tool, cutting this down to slightly above its half-diameter. This countersink may be run a great deal faster than the standard form; it cuts less freely, however.

The Counterbore or Pin Drill

When the worker needs to form a recess for the head of a cheese-head screw he makes use of a counterbore, sometimes called a pin drill. This tool comprises a body having four cutting edges at right angles to one another and a pin or pilot of a size to suit the work in hand. The counterbore is illustrated in *Fig. 10*. The same tool can be used for spot-facing; that is for forming a seat on work upon which may be set nuts or screw heads. For the most part these cutters are machined from a single piece of tool steel. Naturally there is then no interchangeability between the body and the pilot. For this reason spot-face cutters having this interchangeability have been produced for industry.

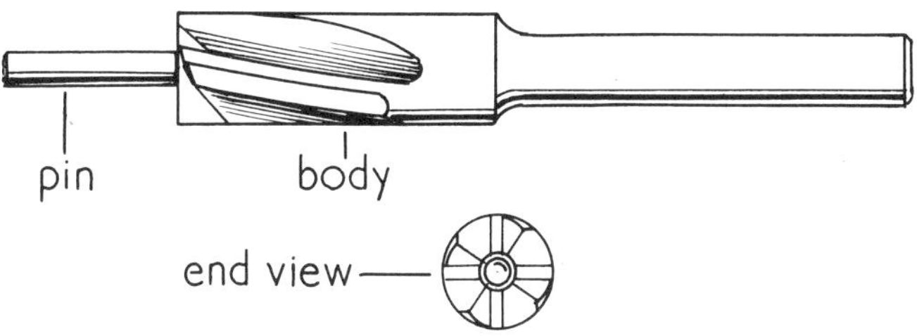

Fig. 10

When modified, a normal twist drill makes a convenient and readily-made counterbore. A group of modified drills is illustrated in *Fig. 11* together with some larger pin drills and other cutters used for counterboring and spot-facing. The modification to the twist drill involves just grinding the end of the drill flat, then backing off the cutting edges so that the drill will again cut freely. The experienced mechanic can readily make the necessary modification; but the inexperienced worker need not be deterred, for a little practice in grinding old drills will soon bring about proficiency.

The two large spot-face cutters depicted in *Fig. 12* were made by the author during the progress of work on a vintage motor car. As will be seen they were of a somewhat large size and were hand-powered.

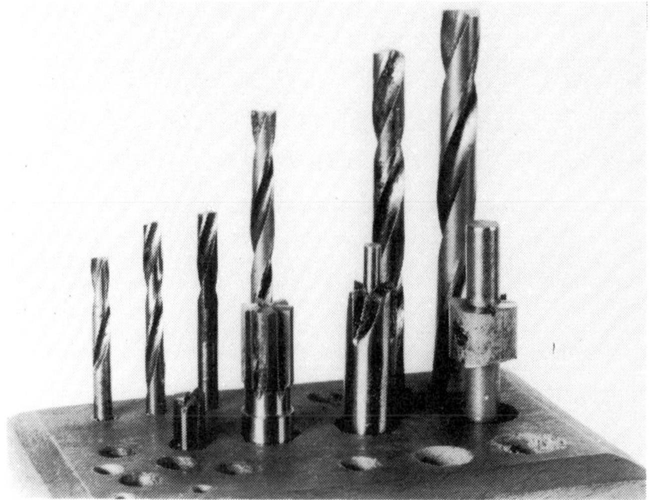

Fig. 11

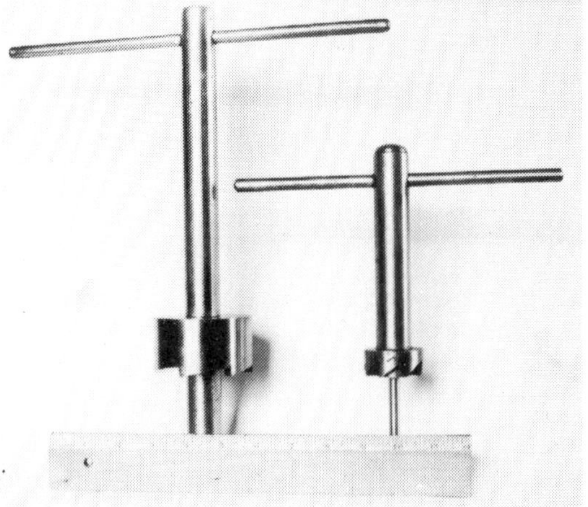

Fig. 12

Hand Drills

While drills and countersinks are employed mainly in the lathe or the
drilling machine, they have sometimes to be used in a hand drill, a small

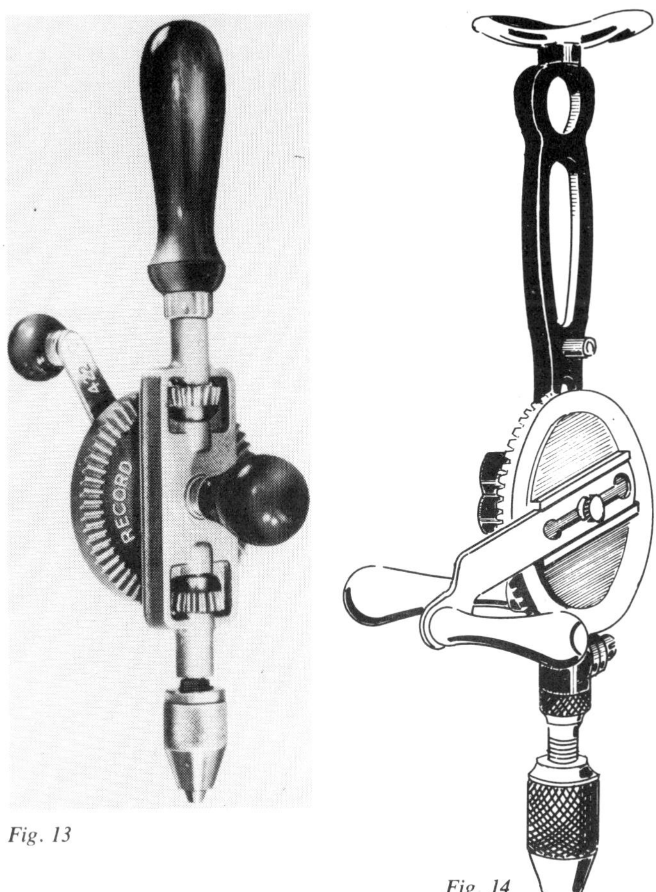

Fig. 13

Fig. 14

version of the tool being illustrated in *Fig. 13*. This is a single-speed machine fitted with a 3-jaw chuck of 3/16 in. or ¼ in. capacity, which is about as large a drill as can comfortably be handled in a hand drill of small capacity.

The drill illustrated in *Fig. 14* is a much larger affair, having a 2-speed system and a chuck capable of accepting drills up to ½ in. diameter. As will be seen the machine is provided with a pad, set at the top of the main frame, against which the user can lean his chest in order to exert sufficient force when feeding the larger sizes of drill to the work. For this reason the tool is sometimes called a breast drill. It appears to be going out of favour and presumably out of production, to be replaced by the electrical hand drill that needs little or no exertion when using it. This seems a pity; for all its sophistication the electric drill is apt to provide an incorrect speed for many sizes of drill, whereas the hand machine allows the user to adjust the drill-speed in accordance with the size of the drill in use. The particular breast drill illustrated was made by the Record Company who have since taken it out of production. It forms part of the equipment in the author's workshop where, despite the power-driven tools available, it often finds employment. It is in all probability the best of its type; it has an adjustable ball-bearing chuck spindle and is provided with a spirit level to help the user when trying to keep the drill on an even keel.

Electric Hand Drills

Modern practice has substituted the electric drill for the hand-operated form. A typical example is illustrated in *Fig. 15*. The same type of drill can be set on a stand and used as a bench drill in the manner depicted by the illustration *Fig. 16*.

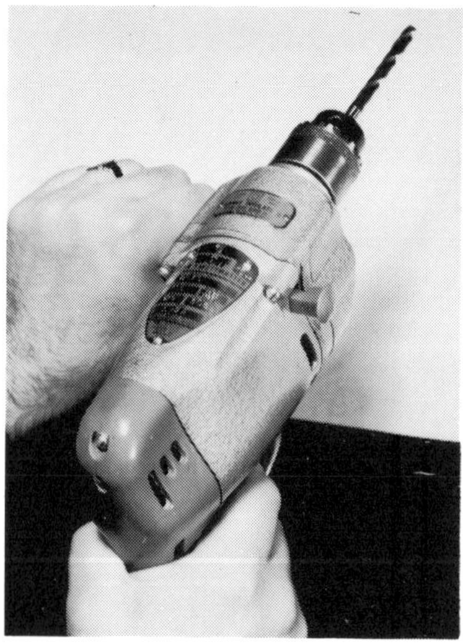

Fig. 15

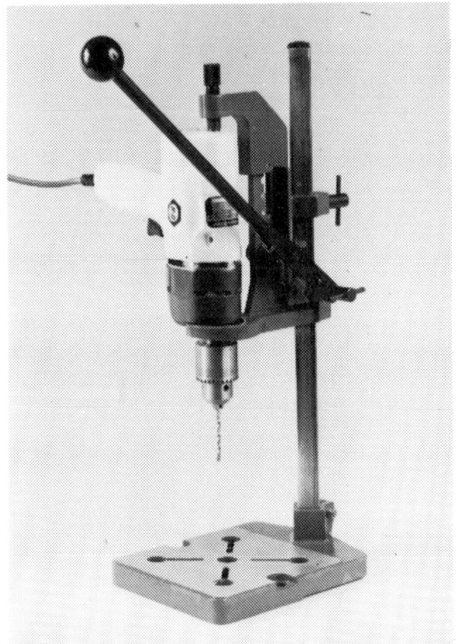

Fig. 16

Hand-operated Bench Drills

At one time a number of hand-operated bench drills was available to the worker. Some of these tools were provided with an automatic feed for the drill, others had manual feed only. While many of these machines were strongly built and capable of drilling large holes, some of them were of light construction and only capable of correspondingly low performance.

Many of the better class of bench drilling machines were to be found in country blacksmith's shop. These tools were fitted with large flywheels in order to store up energy and so make the drilling of large holes a comparatively easy matter.

It is possible such machines still exist so a typical light bench drill is depicted in *Fig. 17*.

Reamers

Although a reasonable standard of accuracy may be expected from holes made by a twist drill, for some purposes neither this accuracy nor the finish imparted to the surface of the holes is good enough. It is then work for the reamer to provide the accuracy and finish needed. For the

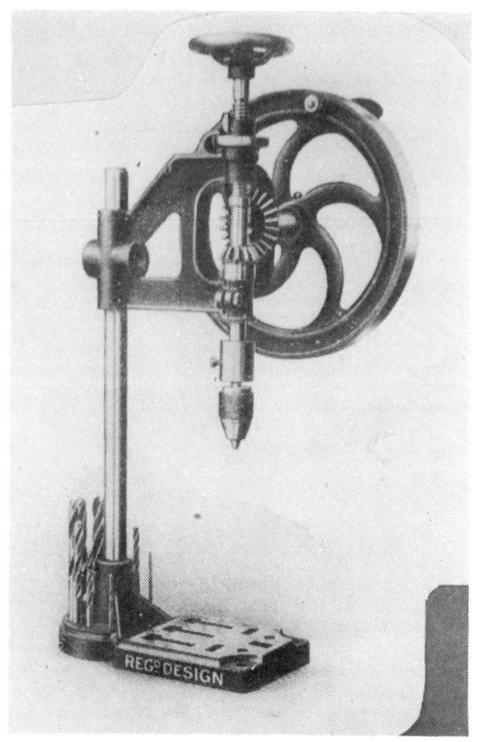

Fig. 17

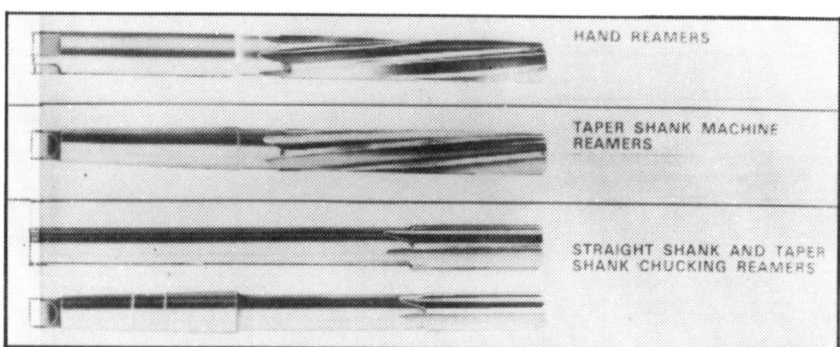

Fig. 18

most part reamers are parallel with a slight taper at the leading end and they are available in the forms set out in *Fig. 18*.

It will be observed that in some cases the blades are parallel with the axis of the shank, and in other cases that they are helical. The latter formation inhibits chatter and so promotes a better finish on the work.

In addition to parallel reamers for use either by hand or in the lathe, there is a number of specialized tools that are applicable to specified reaming operations. The first of these is the adjustable parallel reamer illustrated in *Fig. 19*. The detachable blades of the tool are free to slide up tapered ways machined in the body of the reamer, and are moved and locked, after adjustment, by the barrel-nuts seen at each end of the blades. It will be clear that the action of moving the reamer blades up or down the inclined planes, formed by the tapered ways, serves to expand or contract the range of the adjustable reamer.

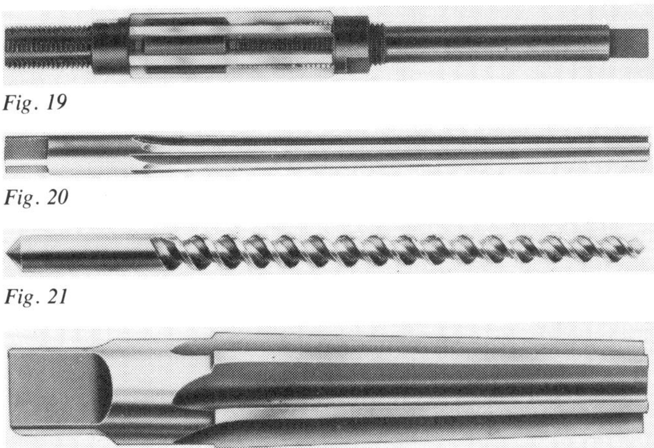

Fig. 19

Fig. 20

Fig. 21

Fig. 22

The tool depicted in *Fig. 20* is a taper pin reamer. Its purpose is to ream the seatings for taper pins that are used to secure components on spindles or shafts. The example illustrated is for hand use. Industrially, hand operation is not acceptable in every case, so the machine taper reamer seen in the illustration *Fig. 21* has been introduced.

Finally, many machines have spindles bored to standard tapers in order to accept tools or equipment having arbors made to these tapers. Sometimes the machines suffer damage to their spindles, in the form of slight tears to the surface of the taper. It is the purpose of the tapered reamer shown in *Fig. 22* to correct the damage, taking light cuts by hand.

CHAPTER V

SCREWING TACKLE

SCREW THREADS CAN be divided into two classes — internal and external. Both can be produced by hand methods using the equipment we shall be discussing.

Internal Threads

Internal threads are cut with taps made from tool steel. The taps are themselves threaded to the pitch required and are fluted, generally with four flutes, sometimes with three, in order to present a cutting edge and to provide an escape route for the chips produced in the process. Taps for hand use have a square machined at one end, as seen at A in *Fig. 1,* so that a tap wrench can be applied. Taps intended for machine use, and in particular those required for the tapping of nuts, have plain shanks so that they may be gripped in a chuck or other holding device.

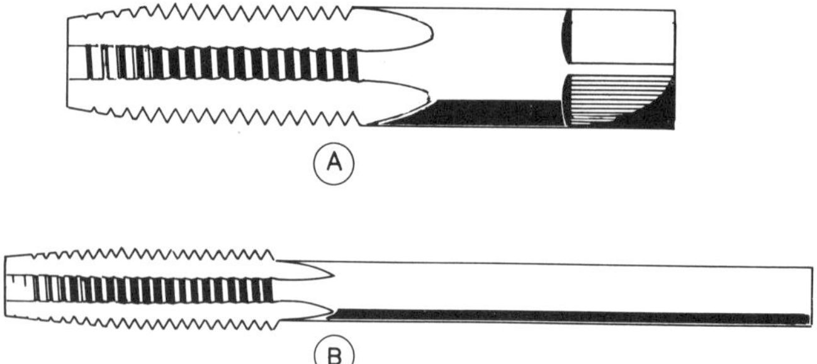

Fig. 1

In the case of machine taps the diameter of the shank is such that, when used for making nuts, the nuts will pass over the shank and congregate on it. A typical machine tap is depicted in *Fig. 1* at B.

A typical example of tapping nuts in a machine is depicted in *Fig. 2*. Here, as will be seen, the nuts are held in a fixture; and, as they are tapped one-by-one, they ride up the shank of the tap, where they collect. As soon as the numbers collected inhibit further tapping, the tap is withdrawn from the chuck and the nuts removed. The tap can then be replaced and the work again commenced.

Fig. 2

Fig. 3

At this point the size of the drilled hole before tapping needs to be mentioned. Clearly, in order that threads shall be formed at all in the work the drilled hole must be smaller than the outside diameter of the tap itself. The size chosen is approximately slightly larger than the core diameter of the tap, that is the measurement taken over the roots of the thread as indicated in the diagram *Fig. 3*.

It follows that the size of the drill chosen will depend to some extent on the material to be tapped; but it also is governed by the extent of engagement between the internal and external components of the thread. This is usually expressed as a 'percentage-of-engagement'.

Hand taps are made in sets of three. These are depicted in the illustration *Fig. 4*. The taper tap is used to start the work; if turned with light endwise pressure it will start, and once engaged will continue to pull itself into the work. The particular example illustrated has a considerable length of parallel lead. This in itself does no cutting; if, therefore, the drill hole allows the lead to enter without shake the tap will start squarely. However, not all taper taps have this provision so to ensure success the worker must check the squareness of the tap as it is entering the work. This is done by applying a small square, placed on the work, to the tap itself in order to test its uprightness.

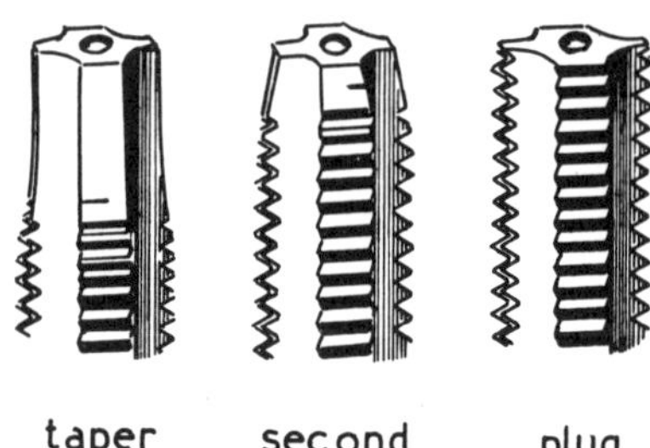

Fig. 4

The 'second' tap, seen at the centre of the group in the illustration, is used after the taper has penetrated the work fully to make sure that the thread so formed is parallel throughout its length. When the hole to be tapped is 'blind' all three taps are used in the sequence illustrated, the plug tap making sure that the threads extend right to the bottom of the hole.

Tap Wrenches

The wrenches used to drive the tap vary considerably in type. Their size and the leverage they exert is usually chosen by the manufacturers

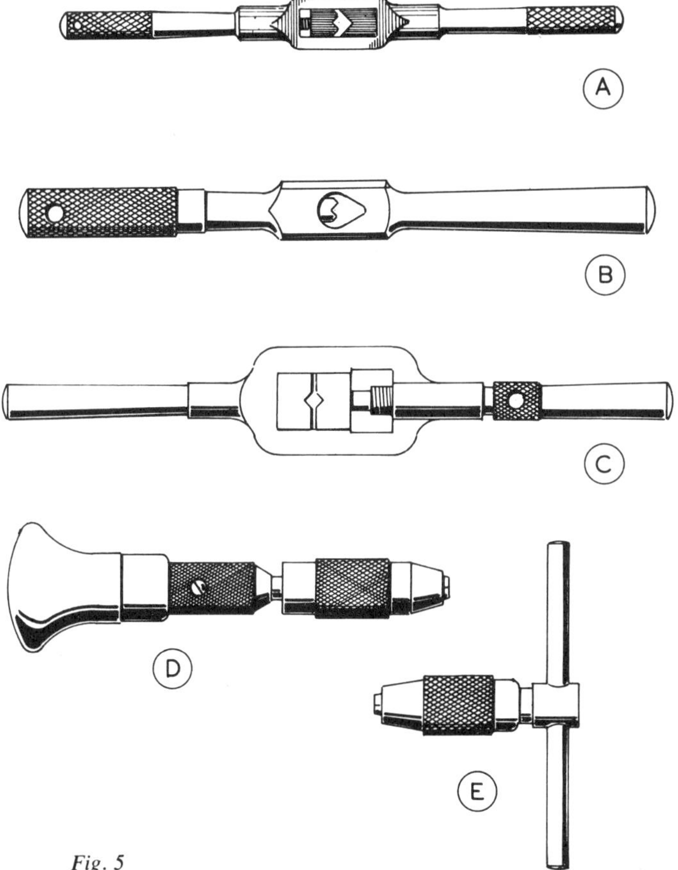

Fig. 5

to suit the range of taps with which they will be used. A representative group of tap wrenches is illustrated in *Fig. 5*.

Of the tap wrenches depicted that at A is of American design and is or was supplied with the 'Little Giant' screwing tackle once very popular.

The wrench at B is also of American design and is the product of S. W. Card of that country whose equipment is of the highest class.

Wrench C is of somewhat historic interest only, as it combines a die-holder and a tap wrench. In these tools the dies, split and in two

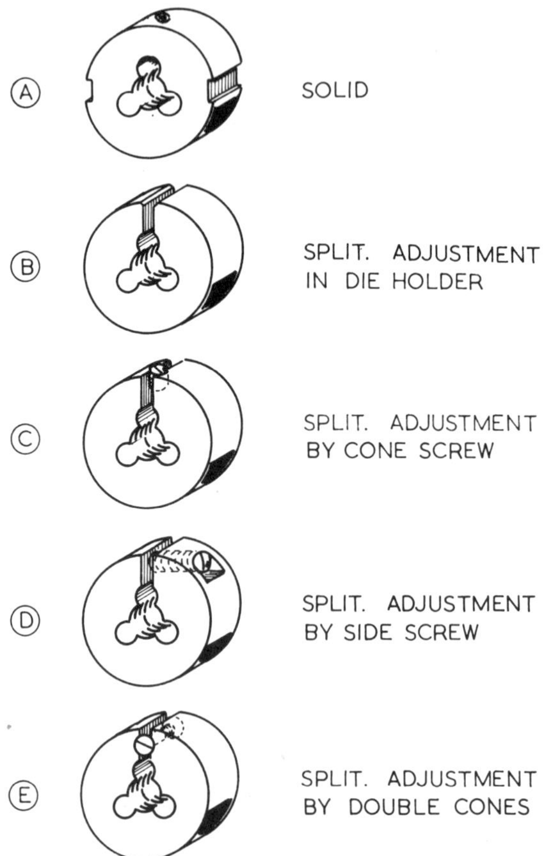

Fig. 6

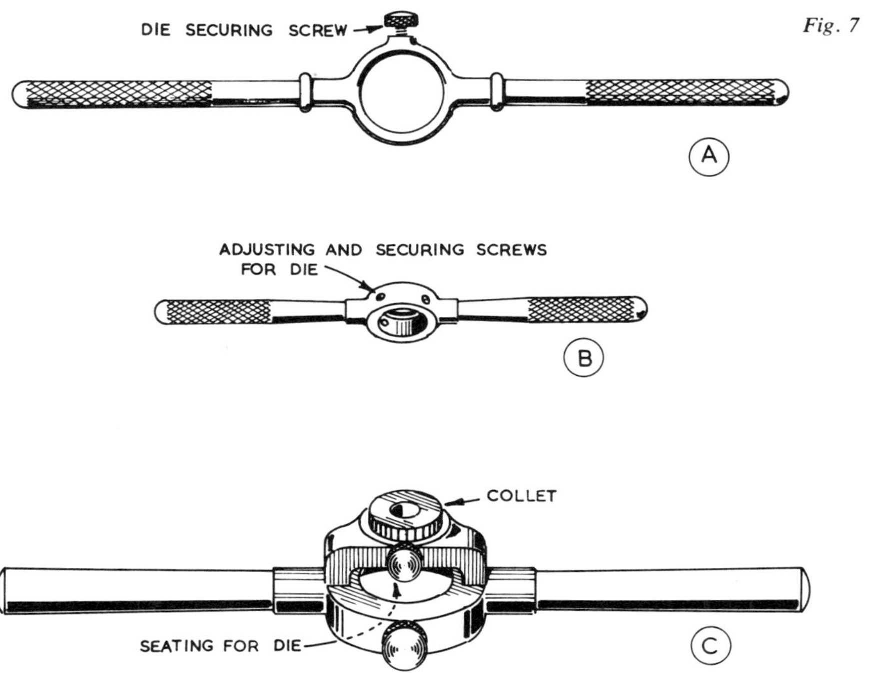

DIE SECURING SCREW
Fig. 7
A
ADJUSTING AND SECURING SCREWS
FOR DIE
B
COLLET
SEATING FOR DIE
C

parts, could be replaced by a pair of blocks as seen in the illustration, the blocks being adapted to engage the square on the shank of the tap.

The tap wrench illustrated at D is intended for small size taps only. It is fitted with a reversible ratchet device that allows the tap to be entered or retracted without continuous or complete rotation of the wrench itself.

The T-handled tap wrench E is of a type made by Moore & Wright of Sheffield. In order to ensure that their leverage is correct for small and medium size taps these wrenches are available in two sizes.

External Threads

External threads are formed by circular or 'button' dies of the type illustrated in *Fig. 6*. As will be apparent these dies vary in many details, most of them concerning the methods sometimes used to adjust these tools so that they will cut the thread correctly.

Modern practice seems to prefer the solid die as these are now made to a high standard of accuracy. The threads are often ground so that they cut freely and to a size in conformity with the ground thread taps that are also available. Button dies are held in holders of the type illustrated in *Fig. 7*. That normally used is the holder depicted at B. This, as will be seen, is fitted with three set screws that serve to secure and adjust a split die when this is in use.

The holder at A is that made for 'The Little Giant' tackle mentioned earlier.

Colleted Die-holders

One of the problems in using button dies is the difficulty in starting them square with the work. 'Little Giant' overcame the difficulty by fitting a pair of threading cutters in a body which itself was provided with a guide to make sure that the thread was started truly and remained so throughout the operation. The combination is illustrated in *Fig. 8*. With such an arrangement no interchangeability is possible, since the guides are of the same nominal diameter as the thread set by the cutters.

S. W. Card's solution to the problem is illustrated at C in *Fig. 7*. From this illustration it will be seen that the collet and the die are completely interchangeable. The collet is simply a bushing drilled concentrically to a definite nominal size. This is fitted to the die-holder in the manner shown. Consequently, therefore, to thread a piece of half-inch material that has been shouldered to a quarter-inch diameter one need only fit a half-inch collet to the die-holder whilst the ¼ in. die is set in the appropriate seating.

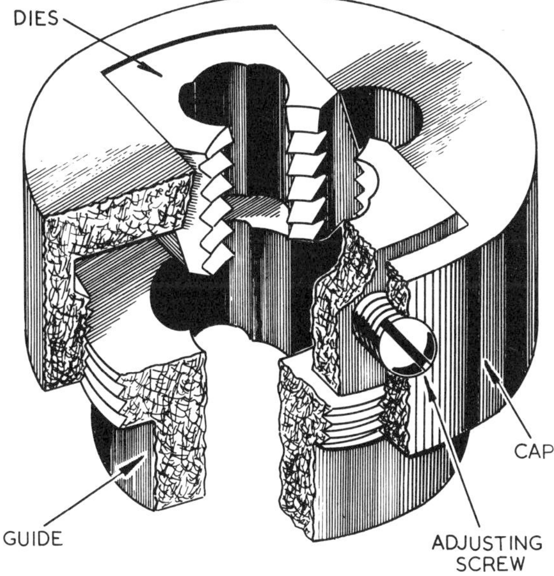

Fig. 8

The illustrations *Fig. 9* and *Fig. 10* show one of a series of colleted die-holders made by the author some years ago and still in frequent use. As to the outside diameter of button dies in general use, in most workshops dies 13/16 in., 1 in. and 1 5/16 in. diameter are largely employed.

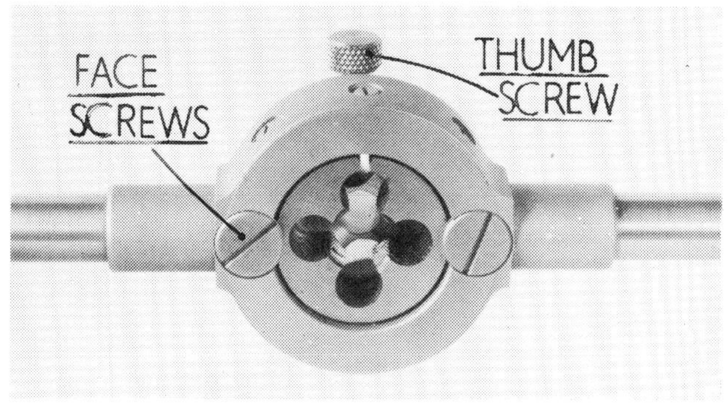

Fig. 9

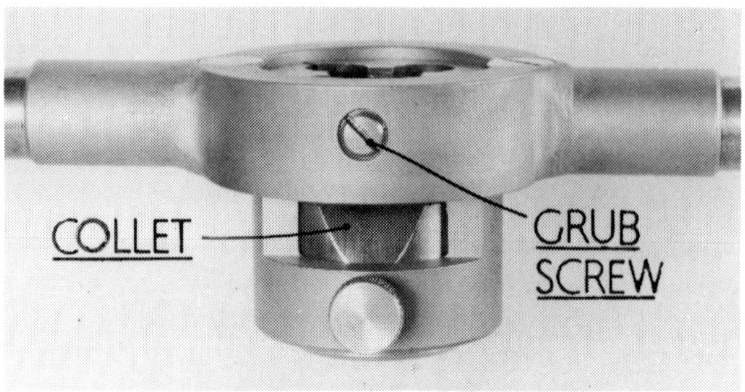

Fig. 10

Die Nuts

One further form of die needs mention; this is the die nut depicted in *Fig. 11*. It resembles an ordinary nut that has been converted into a circular die, and its purpose is to dress damaged threads principally on standing machinery.

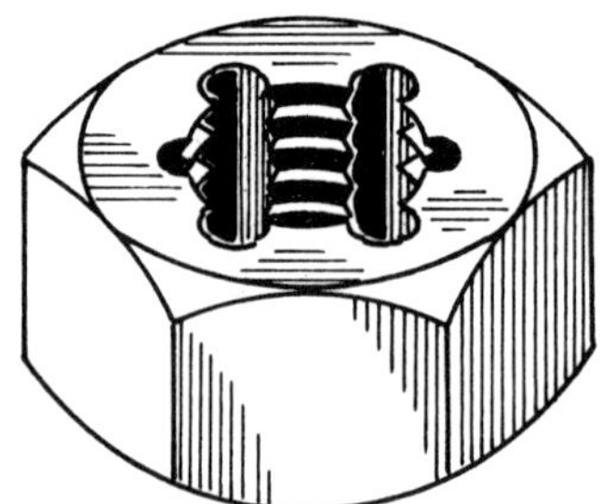

Fig. 11

As to the choice of threads — at the time of writing readers will be aware that there is a move under way to introduce metrication generally. Inevitably, this move will effect screwing tackle in the future; for the moment, however, the potential metal worker might well base his purchase of any equipment to correspond with the nuts and screws he is likely to use in his work. There is a wide range here, a range for which space in this book is not available. Readers who need further information on this matter might care to consult the author's book *Screw Threads and Twist Drills* published by Argus Books.

Mention has already been made of the advisability of using a little oil when cutting either external or internal threads. Whilst engine oil may well serve in mild material, it is better to use one of the special compounds made for the purpose. The author tends to use one that has a sulphur basis, and there is little doubt that a compound of this character materially lessens the labour when cutting threads.

CHAPTER VI

MISCELLANEOUS TOOLS

FINALLY, WE COME to a series of hand tools necessary in any work-shop and without which much work, involving the assembly and dis-mantling of machinery for example, would be impossible. The tools in question are hammers, screwdrivers, spanners and pliers in various forms with which we shall deal in due course.

Hammers

Various hammers find a place in the engineer's workshop, but the most important is the 'ball-pane' hammer illustrated in *Fig. 1*. The ball-pane is the round knob at the back of the hammer head and is generally used for riveting. A hammer weighing some 1½-2 lb is generally considered satisfactory, but a similar type, say ½-¾ lb in weight may be advisable when much light work is to be undertaken.

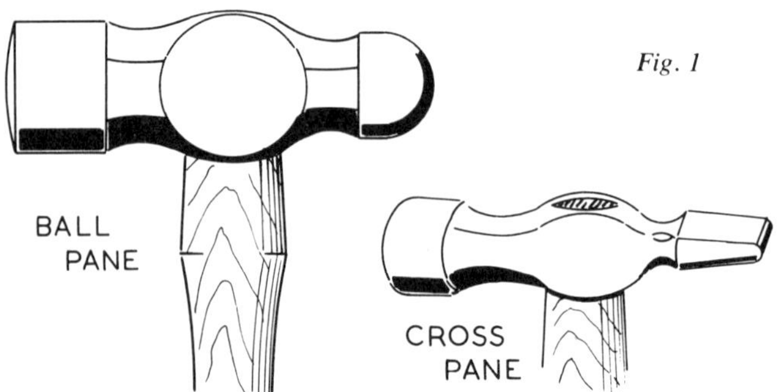

The cross-pane hammer, also seen in the illustration, is sometimes useful in connection with sheet metal work. When hammering finished work a lead hammer, made by attaching or casting a block of the metal onto an iron shaft, will be found of service.

Somewhat more durable, however, is a hammer made from brass. These are made, for the most part by the operative himself, and comprise a short length of brass rod, say 1¼ in. diameter attached to a steel shaft some ⅜ in. diameter fitted with a wooden handle. The arrangement is illustrated in *Fig. 2*. A similar type of hammer, made from rolled-up raw-hide has found favour for many years when dealing with finished or partly finished components.

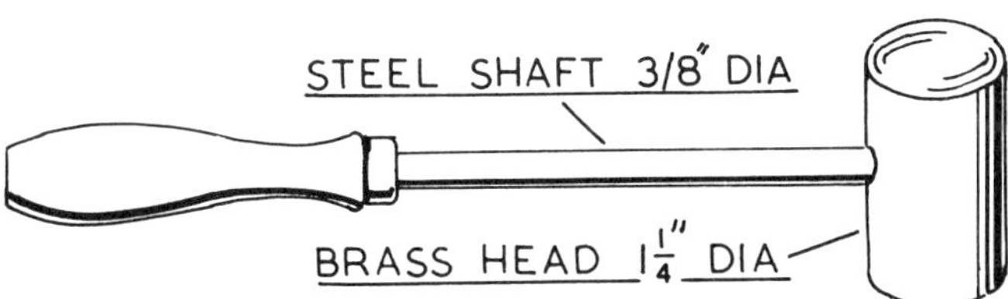

Fig. 2

Screwdrivers

A variety of screwdrivers will be needed in the workshop, the size and number depending on the class of the work that is undertaken. Examination of any reputable tool catalogue will quickly reveal to the reader that the choice is wide.

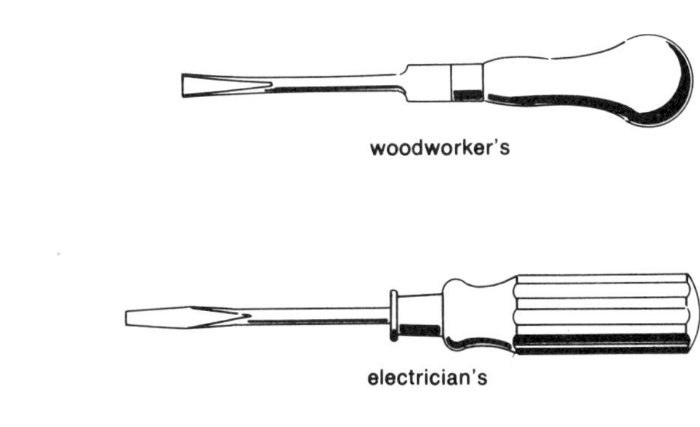

woodworker's

electrician's

Fig. 3

watchmaker's

Basically, there are three types of screwdriver that should be available in the workshop; these are the woodworker's or mechanic's screwdriver, the instrument screwdriver and the screwdriver for the worker indulging in much electrical work. Of the woodworker's type perhaps two or three sizes will be needed. As they vary in length so they differ in blade width, thus enabling the user to choose the right tool for the job.

Instrument screwdrivers, sometimes listed as watchmaker's screwdrivers, vary in blade width but often are all the same length so that they are generally sold in sets.

The electrician's screwdriver is for the most part similar to the mechanic's, but having its blade set in an insulated handle capable of resisting the pressure of the mains electrical supply.

The three types are depicted in the illustration *Fig. 3*.

Spanners

Readers will scarcely need to be told for what purpose spanners are used in a workshop. However, some notes on the variety that are available may be of help to those workers who are in the process of gathering a useful collection of spanners together.

It is advisable to purchase good quality tools that can be guaranteed to fit accurately the nuts with which they are to be used, and are made from material that will stand wear.

The open-ended spanner illustrated in *Fig. 4* is the type most commonly used. Unless the work envisaged dictates otherwise a set with a range from 3/16 in.–½ in. diameter will suffice.

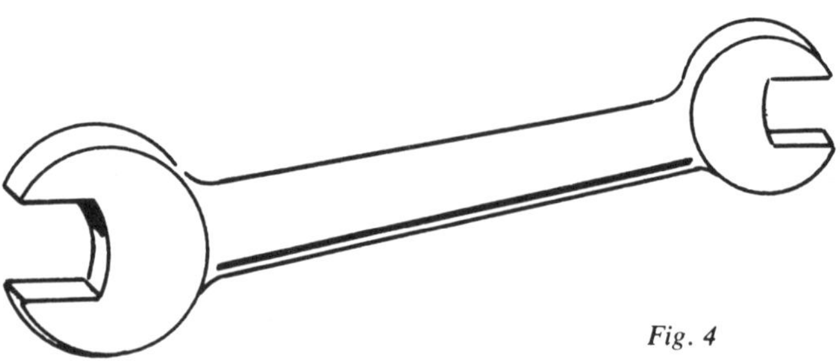

Fig. 4

The same remarks apply to the box spanner depicted in *Fig. 5*. These are of robust construction and will stand up to a lot of hard work.

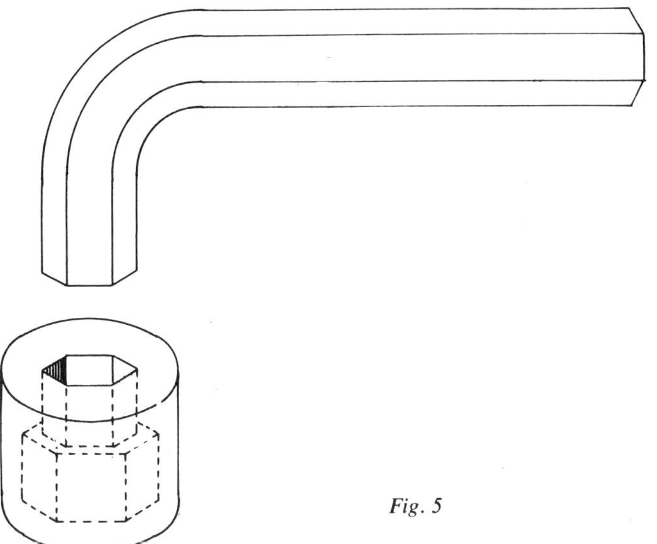

Fig. 5

The tube box spanners (see *Fig. 6*) are for lighter duty. They are also useful, sometimes, when nuts are otherwise in inaccessible places. The products of the motor industry are of course, well known for providing examples of awkwardly placed nuts. To deal with these difficulties specially designed spanners and attachments are obtainable.

Fig. 6

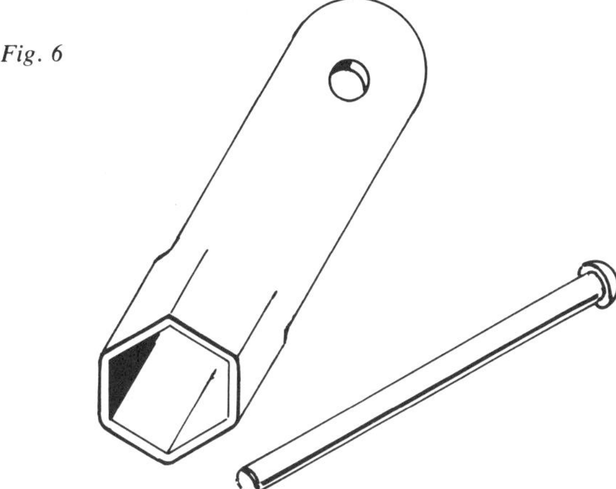

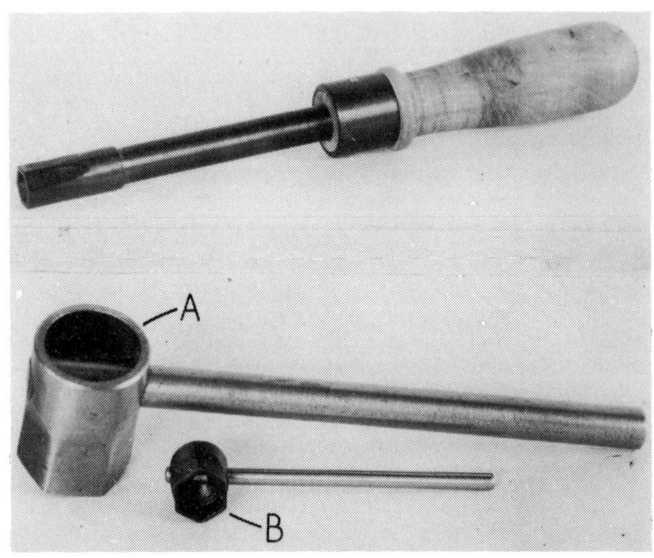

Fig. 7

Fig. 8

Instrument and small electrical work requires the use of purpose-made box spanners of the type illustrated in *Fig. 7*. The particular examples shown were made by the author, but spanners of this type are obtainable commercially. Larger spanners of a similar type are illustrated in *Fig. 8*. They were once available in groups up to ½ in. nut size; if they are still obtainable they are well worth including in a tool kit.

Ring Spanners

Some fifty years ago, in the motor and motor cycle industry, there were many components, such as valve caps and the like, that needed large spanners in the form depicted on the right of the illustration *Fig. 9*. The hexagons machined on these components were sometimes 1½ in.–2 in. across the flats, whilst the flats themselves were usually not more than ½ in. high. The positioning of these components often precluded the use of open-ended spanners without fear of damage. Ring spanners of the type illustrated were then supplied to enable the motorist to remove and replace the parts in question.

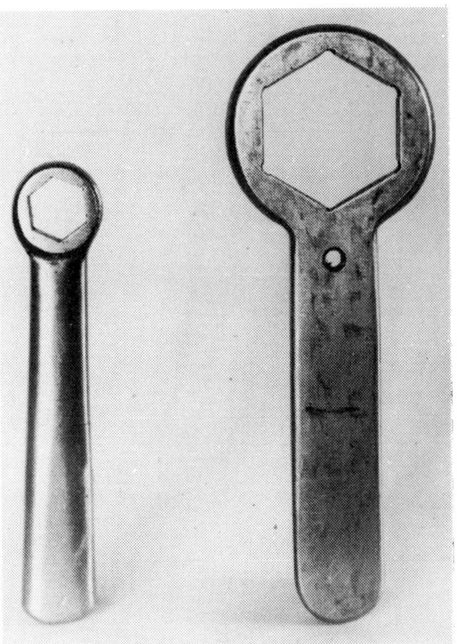

Fig. 9

The small ring spanner, also illustrated, sometimes finds a use around machine tools. In this context the author sometimes makes use of them—on nuts or hexagon-head screws that otherwise would need the frequent use of an open-ended spanner. As these particular spanners are specially made for the most part they are normally rendered captive by fitting a grub screw to secure them. An example is depicted in *Fig. 10*.

Fig. 10

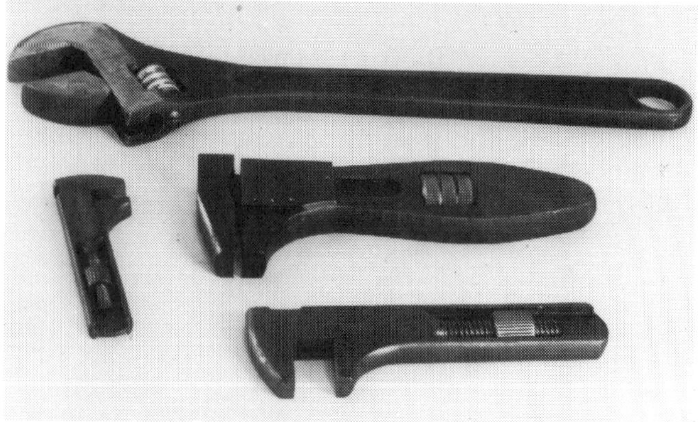

Fig. 11

Adjustable Spanners

A representative group of adjustable spanners is illustrated in *Fig. 11*. They have the advantage that they can get the user out of trouble when no fixed-jaw spanner of the right size is to hand.

If carefully used and set to fit the nut firmly they will do little damage to its facets. On the other hand when adjusted incorrectly they are apt to bruise the corners of the hexagon more than somewhat. The reader, therefore, should conclude from these remarks that a kit of spanners consisting of the adjustable type only is not to be recommended.

Pliers

Pliers are used for many duties in the workshop. Whilst they may be purchased individually, each one being designed for the particular duty required of it, one may condense the newcomer's needs, at least initially, to the three basic forms depicted in *Fig. 12*. First we have the combination plier. As its name implies it combines in one tool two

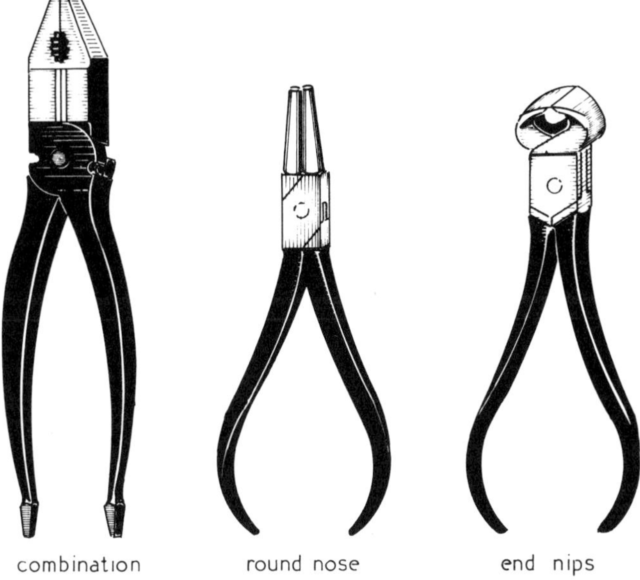

Fig. 12

forms of wire-cutter, a pair of serrated jaws for work on gas fittings and a flat nose pair of jaws for purposes that will no doubt suggest themselves to the reader.

The round nose pliers seen at the centre of the illustration are virtually exclusively used for wire work, forming eyes and the like. Employed with some dexterity, albeit with some difficulty, they may be used to open circlips when there is need to remove them from some piece of mechanism or other.

Finally the end nips. These are essentially a wireman's tool, be the wireman electrical or otherwise, to be used for cutting plain wire or electrical cable and stripping off the insulating covering when needed.

Examination of any reputable tool merchant's catalogue will show that in addition to the three basic pairs of pliers already described and illustrated, there are numerous other forms that may be of service in the workshop at some time or other. Amongst these are the gas pliers, sometimes called burner pliers, illustrated in *Fig. 13* and the extension flat nose pliers seen in *Fig. 14*. The former, as its name implies, was designed primarily to deal with small gas fittings but is equally useful when any round material within its capacity needs to be handled. The flat nose pliers are principally of service when, for example, during an assembly operation nuts in an awkward location need to be held.

Fig. 13

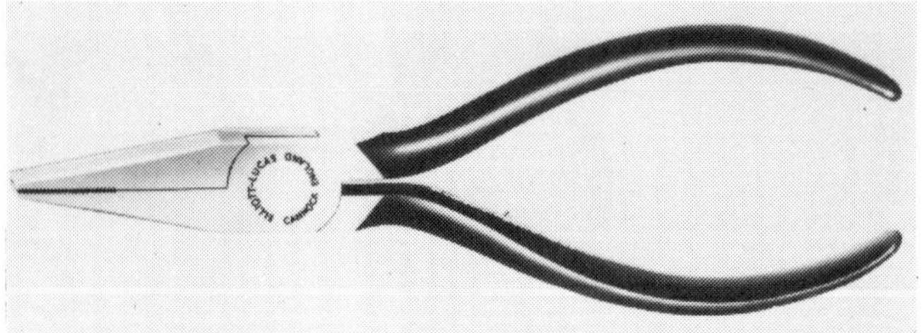

Fig. 14

Scrapers

When the removal of small quantities of metal is needed, either to correct a fault or to ensure the accurate fitting of some component, the mechanic makes use of a scraper.

Examples of the class of work involved are the fitting of bearing brasses and the scraping of two flat mating surfaces in order to ensure their congruence. For the scraping of bearing surfaces use either the triangular scraper or the bearing scraper illustrated in *Fig. 15* and *Fig. 16* respectively. These tools may be purchased; on the other hand, many workers make them for themselves, for the most part from old files.

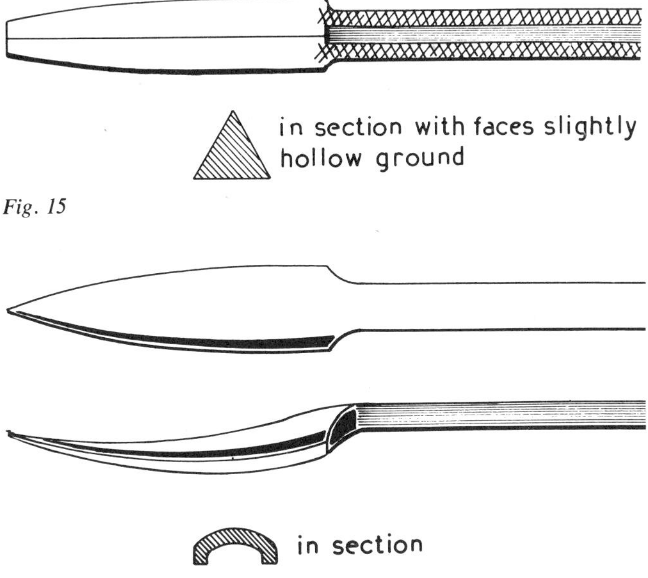

Fig. 15

Fig. 16

For the scraping of flat surfaces, on machine tools for example, the hand scraper depicted in *Fig. 17* is employed. Here again the mechanic generally makes them from an old flat file.

It should be added that the flat scraper is also used to impart a decorative finish to work when needed. This is known as 'frosting' and is produced by a series of overlapping cuts at right angles to one another

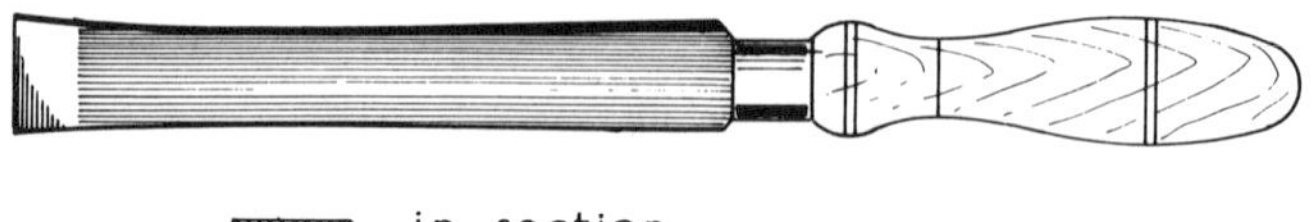

Fig. 17

in the manner portrayed by the diagram *Fig. 18*. The particular details of the operation have been described elsewhere.

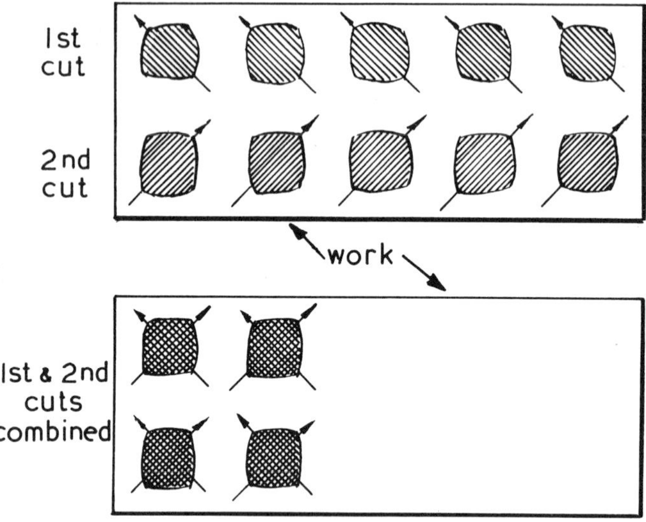

Fig. 18

All scraping work depends for its success on the correct preparation and sharpening of the tools with which it is carried out. Space does not permit touching on this matter here; but those readers who may need further information will find it in *Sharpening Small Tools* published by Argus Books.

Grinding Machines

Having touched on the matter of tool sharpening, this would seem an opportune moment to consider the equipment available for the purpose. The sharpening of tools is primarily a grinding operation carried out by abrasive wheels that are either hand or power driven. At one time the only abrasive material available was natural emery and sandstone. The former could be made into wheels for grinding engineers' cutting tools whilst the sandstone, also made up as a wheel, was used to sharpen the chisels and planes used by woodworkers.

To-day the demand is for artificial abrasives that sharpen tools quickly and these are now made into wheels that are fitted to the electric grinders now on the market.

Modern electric grinders are fitted with a pair of wheels, one coarse and one medium fine, able to deal with the usual run of tools to be found in the metal working shop. The coarseness of a grinding wheel is defined by a number that has reference to the size of the abrasive grains of which the wheel is composed; the higher the figure the finer is the grain structure. Most grinders are fitted with one 60 grit and one 80 grit wheel, 'grit' being the term applied to the grain size.

A typical electric bench grinder is illustrated in *Fig. 19.*

In addition to equipment just mentioned the workshop will, of course, need at least a couple of hand stones, one coarse and one fine, in order to service such tools as scrapers and any tools for woodwork the workshop requires.

A further word on the subject of grindstones; as these machines normally have sandstones mounted on the spindle they are principally of interest to woodworkers whose plane irons and chisels are best served by these slow-cutting water-immersed abrasive stones which will not draw the tools' temper. New equipment of this class is expensive and not of any general use to the true metal worker so he is not likely to spend much money on it.

Nevertheless, if a good second-hand example should come to the notice of any interested reader, it might be worth his while to invest a little money in order to secure it; if only to service the few carpenter's tools he will need in the workshop.

Sandstone needs to be treated with some care. In the first place a sandstone wheel should never be allowed to remain in water once the grinding operation has been completed. Care should also be taken to see that the work is not allowed to dwell at any one point on the periphery of the wheel; remembering that sandstone is relatively soft and develops a groove easily if the work is not moved continuously across the face of the wheel.

Driving the Grindstone

Three methods of driving the stone are available, namely power, treadle and hand. Remembering that, in order to grind tools successfully, the

Fig. 19

Wood Frame Grindstones.
Iron Frame Grindstones.
OUR WATCHWORD
QUALITY
R. MELHUISH LD
Mounted Grindstones.
Best Blue Bilston Grit, specially suitable for Engineers, Carpenters, Butchers, and others.
Connecting Rods and Treadles of Wrought Iron, as are the Spindles.
All Grindstones sent Carriage Forward.
No. 1795.

Fig. 20

use of both hands is essential, the amateur mechanic will opt for treadle driving unless he can press an assistant into service to turn the handle for him. One would imagine that power drive is for the professional only and not for any occasional user. Typical grindstones are depicted in *Fig. 20.*

Anvils

The anvil is used for such work as the straightening out of material, for forging and as a bolster for any riveting operations that may occur. Anvils come in many sizes and weights, so the choice will depend upon

Fig. 21

Fig. 22

the class of work likely to present itself. The small jeweller's anvil, seen in *Fig. 21*, has often been of use in the author's workshop when dealing with small components; whilst the larger unit depicted mounted on a hardwood block in *Fig. 22* comes in for light forging and riveting.

CHAPTER VII

SOLDERING AND BRAZING

Blowlamps

A SOURCE OF heat is essential to the metal worker, not only for the soft and hard soldering operations he will need to carry out, but also for any hardening, tempering and light forging work that may come his way.

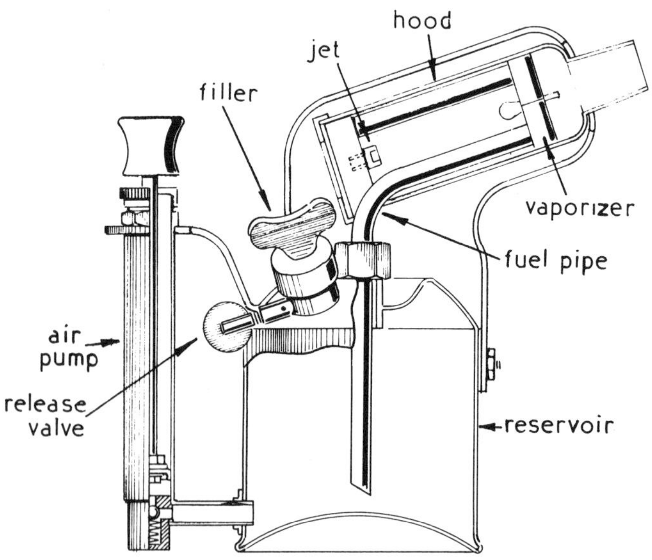

Fig. 1

Fig. 2

At one time, depending on whether the workshop was in the country or in the town, the fuel needed to produce the heat was either paraffin oil or coal gas from the town mains. The paraffin was burnt in a blow lamp of a type shown sectionally in *Fig. 1*, Whilst a representative collection of the lamps is seen in *Fig. 2*.

The working of a blowlamp may be understood from this sectional drawing. The paraffin oil is held in the container which is pressurized by the air pump after the release valve has been closed.

The air-pressure then forces the oil up the fuel pipe to the vaporizer, where after heating it is converted into a gas which is discharged through a jet and is mixed with air inside the hood. The vaporizer, of course, needs to be heated initially. This is effected by placing a small quantity of methylated spirit in the well, seen at the top of the reservoir and lighting it. The heat engendered while the spirit is burning is sufficient to heat the reservoir and start the lamp itself burning, after which the flame is self-supporting since a portion of it inside the hood plays on the vaporizer and keeps it hot.

Blowlamps need to be kept scrupulously clean, the jet pricked clear with the small wire tool provided for the purpose.

Blowlamps are, or were, obtainable in several sizes, for the most part designated by the capacity of the reservoir. Thus the sectional drawing shows a ½ pint or 1 pint lamp. Whilst the illustration of the three in line demonstrates a 1 pint, a 2½ pint and a 5 pint blowlamp respectively. Small lamps seldom have a control on the fuel supplied to the vaporizer; conversely the larger units are fitted with a needle valve to control the fuel and thus the intensity of the flame produced.

Bottled Gas

For some years now gas under pressure has been available to the metal worker. This is a liquid hydrocarbon gas, not town gas, of a high calorific value capable of liquification at a low-to-medium pressure. From the user's point of view the equipment obtainable is very convenient and versatile and far easier to use than the blowlamp. The gas is stored in steel cylinders, which are marketed according to the weight of the gas they contain when fully charged. For example the gas supply in the author's workshop is stored in an 80 lb. cylinder.

Torches used with bottled gas by the author are operated on either high or low pressure according to the type of burner involved. Apparatus for both pressures is illustrated in *Fig. 3*. The torch on the left is for direct working at high pressure whilst that on the right needs a secondary air supply to operate it. Reducing valves for both pressures form part of the equipment and need to be fitted to the gas cylinder before connection is made to either of the torches.

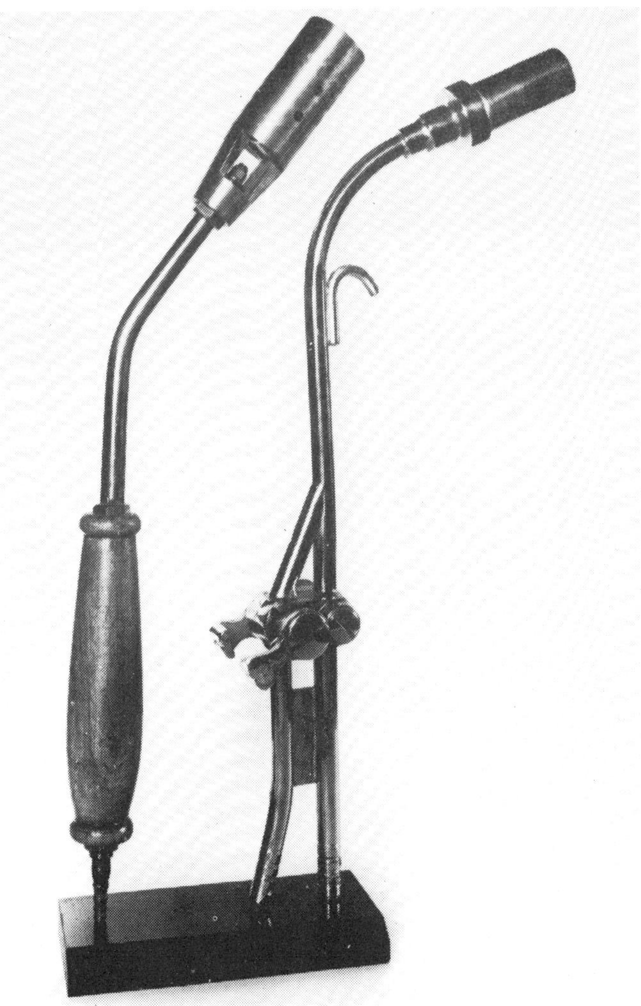

Fig. 3

The torch on the right of the illustration needs an independent air supply. In the author's case this is obtained from the workshop compressed air supply by way of a reducing and regulating valve.

Town Gas

The portability of bottled gas equipment is a genuine advantage to the
country worker, but may not be of much importance to the mechanic
whose workshop is in the town, where like as not a supply of coal gas or
natural gas may be available.

The simplest piece of equipment for use with gas is the Bunsen
burner depicted in *Fig. 4*. It can be used for a number of purposes
including the heating of soldering irons as seen in the illustration.

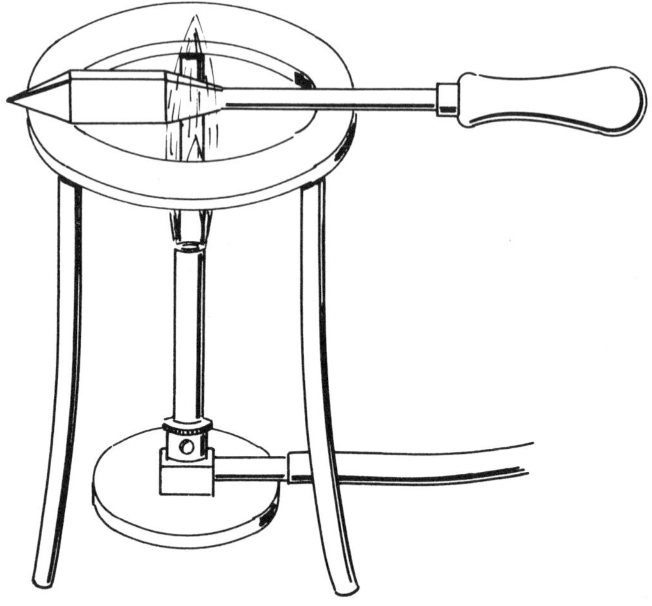

Fig. 4

Brazing torches for town gas and needing a separate air supply are
similar to the torch shown in *Fig. 3*.

In view of the advances that have been made of late years in equipment for use with bottled gas and town gas, readers contemplating their
use would be well advised to consult the manufacturers of the equipment and the gas authorities concerned in order to ascertain the latest
position in this matter.

Soldering Equipment

The metal worker will need a pair of soldering irons. The simplest of
these, a straight and a hatchet bit, are illustrated in *Fig. 5* at A and B
respectively. These bits made from copper affixed to an iron shaft
provided with a wooden handle, are 'tinned' with solder so as to pro-
mote heat transfer between the iron and the work. In this way the
solder will flow freely and produce a neat joint.

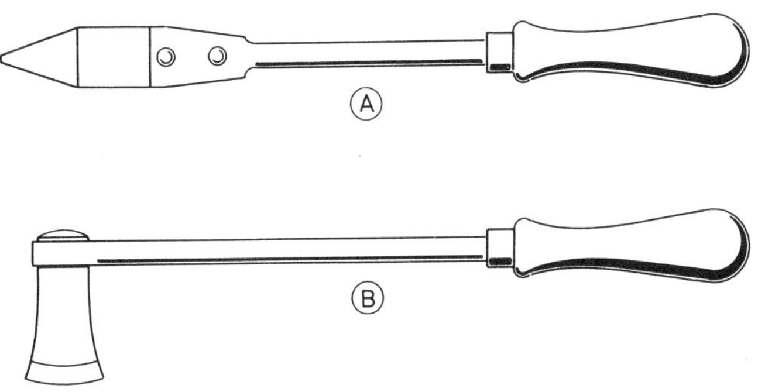

Fig. 5

For those who need something more expensive there are always
electric soldering irons, and even solder bit attachments for bottled gas
torches.

For the most part electric soldering irons are used when soldering
radio components. In this connection the use of resin-cored solder must
be mentioned. This is a solder having as its name implies, a core filled
with natural resin which will flux the work without fear of the joint
being contaminated by corrosion—an essential condition in radio work.

The electrically heated soldering iron is a convenient tool, but not for
heavy work when the weight of a powerful iron somewhat negates its
usefulness. For this reason the professional tinsmith would prefer a
standard iron unencumbered by a somewhat rigid electrical lead, which
plastic instead of rubber insulation has done nothing to improve.

Brazing Hearths

In order to carry out brazing operations some form of hearth is essential
in order to set up the work and surround it with material that will
conserve heat. The same hearth will also serve for light forging and for

Fig. 6

'sweating', which may loosely be described as soldering without a soldering iron, the necessary heat being obtained directly from some form of torch.

A typical brazing hearth is depicted in *Fig. 6*. Here a metal stand supports a tray in which are placed firebricks; these form a work surface whilst the firebrick cubes, also seen in the illustration, are used to surround the work and keep in the heat.

Some form of clamp to hold the brazing torch is advisable; in this way the hands will be left free to carry out the brazing or sweating operations for which the equipment is intended.

INDEX